Mining Stamp Mills

by Algernon DelMar

with an introduction by Kerby Jackson

Introduction

It has been years since Algernon Del Mar released his important publication "Stamp Milling". First released in 1912, this important volume has now been out of print for years and has been unavailable to the mining community since those days, with the exception of expensive original collector's copies and poorly produced digital editions.

It has often been said that "*gold is where you find it*", but even beginning prospectors understand that their chances for finding something of value in the earth or in the streams of the Golden West are dramatically increased by going back to those places where gold and other minerals were once mined by our forerunners. Despite this, much of the contemporary information on local mining history that is currently available is mostly a result of mere local folklore and persistent rumors of major strikes, the details and facts of which, have long been distorted. Long gone are the old timers and with them, the days of first hand knowledge of the mines of the area and how they operated. Also long gone are most of their notes, their assay reports, their mine maps and personal scrapbooks, along with most of the surveys and reports that were performed for them by private and government geologists. Even published books such as this one are often retired to the local landfill or backyard burn pile by the descendents of those old timers and disappear at an alarming rate. Despite the fact that we live in the so-called "Information Age" where information is supposedly only the push of a button on a keyboard away, true insight into mining properties remains illusive and hard to come by, even to those of us who seek out this sort of information as if our lives depend upon it. Without this type of information readily available to the average independent miner, there is little hope that our metal mining industry will ever recover.

This important volume and others like it, are being presented in their entirety again, in the hope that the average prospector will no longer stumble through the overgrown hills and the tailing strewn creeks without being well informed enough to have a chance to succeed at his ventures.

Kerby Jackson
Josephine County, Oregon
May 2016

CONTENTS

CHAPTER V

CHAPTER VI

CHAPTER VII

STAMP MILLING

CHAPTER I

EVOLUTION OF THE GRAVITY STAMP MILL

It is a remarkable circumstance that the ancients who were in many respects first-rate engineers should have overlooked the possibilities of a mechanical device for crushing gold ores. It cannot be that they were lacking in the necessary engineering skill, but rather that they could not see the advantage of combining a water wheel with the pestles already familiar to their workmen and thus save the labor and feed of many men. The sole reason must have been that while slave labor was possible, machinery was not necessary, but when such labor was no longer available or had become expensive, the crude gravity stamp mill used in the fourteenth and fifteenth centuries was adopted from methods and appliances long known to the ancients.

The system of gold milling practised in Egypt in ancient times will be described to show that the next step in the evolution of the gravity stamp mill was easily attainable and required no great amount of adaptability. The quotation is from Doidorus Siculus: "When the stone containing the gold is hard, the Egyptians soften it by the application of fire and when it has been reduced to such a soft state that it yields to moderate labor several thousand of these unfortunate people treat it with iron picks. Those who are above thirty years of age are employed to pound pieces of the stone of certain dimensions with iron pestles in stone mortars until reduced to the size of a bean. The whole is transferred to women and old men who put it into mills arranged in a row, two or three persons being employed at the same mill, and it is pounded until reduced to powder." (Here we have a gravity stamp mill operated by three persons lifting a heavy pestle of wood shod with iron.) " At length the masters took the stone thus ground to powder and carried it away to

1

undergo the final process." The final process was "to wash on inclined tables furnished with two cisterns, all built of stone." Here we have what corresponds to a modern canvas plant using stone floors instead of wood covered with canvas. Evidently the amount of ore crushed was too great for the slaves

Fig. 1.—Ancient stamp mill. (*Agricola.*)

to pan, so it was run over stone tables with a stream of water. It is very likely that amalgamation in some form was practised, for flasks containing mercury have been found in the ancient ruins of Egypt.

Although we have no description of a gravity stamp mill being used by the Romans, it would have been only a short step in evolution to have used mechanical power to lift the heavy pestles of the Egyptians.

We now skip many centuries until the fourteenth century of our era. Mr. T. K. Rose says "It seems certain that in 1340 a stamp mill, used in making gunpowder existed in Augsburg, and that Conrad Harscher, of Nuremburg, owned one in 1435." This strengthens Mr. Bennett Brough's theory that the first stamp mills were used in the manufacture of gunpowder.

The stamp mill was used in Germany in the fifteenth century and in the sixteenth we have the first record of them being used in the crushing of gold ores. From the Hungarian mines are remains that date from the fifteenth or sixteenth century.

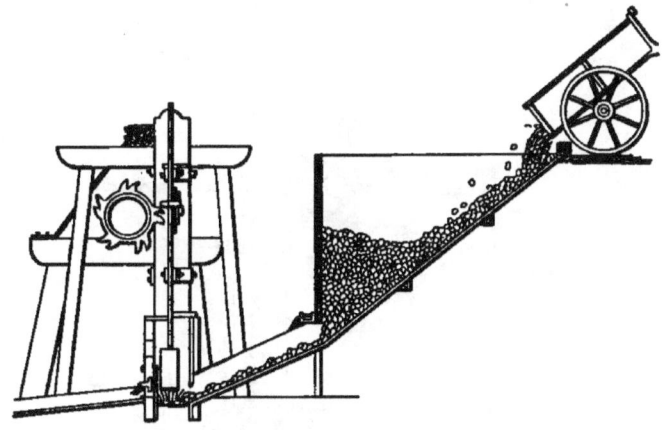

FIG. 2.—Cornish stamp mill. (*Henderson.*)

These mills were constructed substantially as described by Agricola, an illustration of which appears on another page. The tilt-hammer used in China and described by Prof. Henry Louis can hardly be considered as a stamp mill.

The stamp mill as described by Agricola was in use in Cornwall until very recent times. An illustration is here shown. The stamp mill as described by Agricola, when compared with the modern mill, was a very crude affair. The stamps weighed from 150 to 250 lb. and were dropped 8 to 12 in. about seventy times a minute. The cams and tappets were of wood and the stamps were lifted without turning, in pairs when the battery was composed of four stamps and alternately when of three stamps. The cast iron shoe was stuck into the wooden stem and wedged, the stem having rings of iron where the shoe entered to prevent the wood from splitting. A stamp mill of this description was in use in the mines of Potosi, South America, in the sixteenth century, from which date begins the history of stamp milling in America.

From the sixteenth century to the exploitation of the gold fields of California little improvement was made in the gravity stamp mill. In the gold mines of the Southern States of the United States a crude form of mill was in use in the early years of the nineteenth century which was the type of the first mills operated

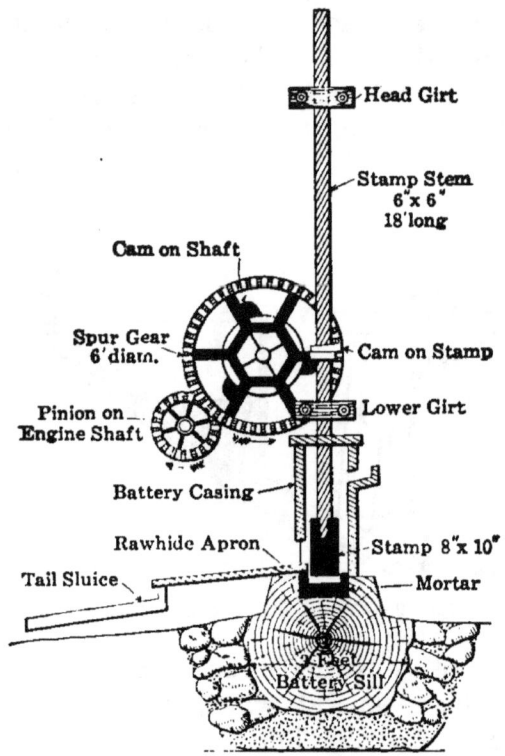

FIG. 3.—Earliest California mill. (*W. S. Moses.*)

in California. About 1835 the first stamp mill in the United States was started at Tellurium, Virginia.[1] The stamps weighed 50 lb. and were made of square wooden stems with iron shoes and dies, the cams operating in slots in the stems.

Fig. 3 shows one of the first mills built in California, erected by Wm. S. Moses and represents the form of construction used in the State of Georgia, from which state all the iron work of the mill was imported.

No great advance was made on the form of mill used in the Appalachian States of America until C. P. Stanford introduced

[1] Crane's "Gold and Silver."

the round stem from which followed the round shoe, die and boss-head; J. Fish suggested revolving the stem to equalize the wear on shoe and die; J. Wheeler and H. B. Angel invented the method of holding the tappet on the stem by means of gib and key and J. M. Scott constructed the double armed cam. These were all inventions of men operating in California and proclaim the modern stamp mill, the California Mill.

The first types of California mills erected were usually under 750 lb. All the stamps were operated from a single cam shaft whether the mill was of four or twenty stamps. Four, five and six stamps were in use as mortar units, but never so far as we know were ten stamps used in one mortar as reported to be in use in Germany. The battery of five stamps dropping in that sequence that gives the best flow of pulp is of late origin.

The modern stamp mills of the best types as used in various parts of the world have been made possible by the improvements in the materials of which the different parts are constructed, due to the constant demand for lower milling costs, and not as in the case of most other classes of machinery where the improvements are mainly in the motor power.

It was the discovery of gold in California and the high cost of labor, that made necessary improved machinery for crushing ores and from this necessity the old gravity stamp mill which is the simplest form of crushing device, was improved until we have it in its present state of perfection. The original mills as well as the hand mortars of the ancients were used simply to break ore that had already been subjected to a preliminary process of comminution. It is a curious fact that the stamp mill is gradually reverting to its original use as a crusher of rock of medium size.

When the cyanide process for the extraction of gold was unknown or in its infancy, stamp mills were in general use as amalgamators, for it was found that a thorough admixture of mercury with the ore in the mortar saved more gold than when the processes were conducted separately. With the introduction of the cyanide process and the finer grinding of the ore, amalgamation to a great extent has been discarded, so that the gravity stamp mill is gradually occupying the place of an intermediate crusher between the rock breaker and the fine grinding machinery. As a coarse or fine crusher it is excelled by other machines, but as occupying the place indicated above it has no peer.

CHAPTER II

GENERAL PRINCIPLES OF THE STAMP MILL

The general principles of construction and operation of the stamp mill is necessarily one of the most important studies for those young men starting on a mining career as well as for others more intimately connected with mining the precious metals. The majority of mining problems that will come before those just starting in the mining business will be problems requiring a great amount of personal work, for these will be propositions where the ore is either low grade or refractory. The treatment of low grade refractory ores will be the main work and to efficiently cover the field all the various details of milling must be followed closely and systematically, and wherever possible a saving made, however small. The work performed by the mill must be carefully checked and every operation conducted under the most favorable circumstances. The crude methods of hit and miss so long in vogue must be given up for more refined methods of analysis both chemical and mechanical.

Sampling

All mill samples should be taken automatically and not left to the caprice or willfulness of the millman. They are but human

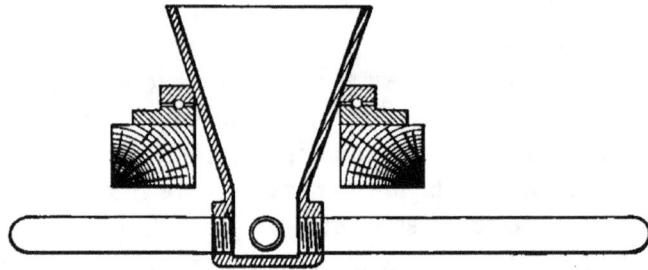

Fig. 4.—Radial sampler.

and are liable to take samples that will show their mill work in the best light. It is possible by simply rejecting the lighter or heavier portions of the tailing sample to vary the results so as to

nullify the whole system of sampling. Two sampling machines
are here illustrated. Both are simple, and do their work well.
The first can be bought at a reasonable price and the second
can be made in a short time with a few boards and a piece of
sheet iron. Fig. 4 is a truncated cone turning on ball-bearings
either at the sides as shown or on the bottom. It has arms of
various lengths curved as familiar to us in the old garden sprink-
ler. The exit from the long arm revolves over a slotted receptacle, the other arms simply exit into a sluice. The impulse of
the tailings flowing through the arms cause the cone to revolve.

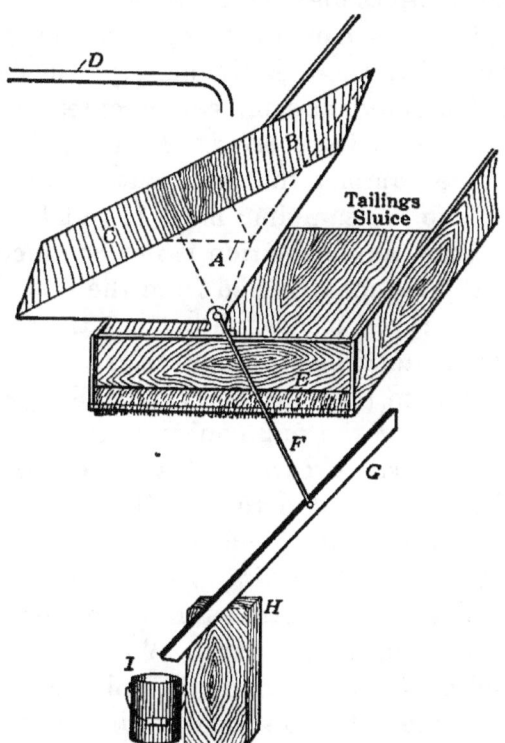

FIG. 5.—Bunker Hill mine sampler.

Fig. 5 shows a sampler at use at the Bunker Hill Mill, Amador
County, California, and is a type of the perfect sampler for it
takes a sample at regular intervals clear across the stream and
the full depth of the stream.

The triangular prism A is of wood or tin, and divided into the
compartments B and C. It is pivoted above the tailing discharge E. The rod F is rigidly attached to A so it will move the
slotted tin conveyor G back and forth across the stream of

tailings. G is pivoted on the block H and discharges into the bucket I.

The operation of the machine is as follows: The jet of water which regulates the frequency of the samplings comes through the pipe D, and when one of the compartments C or B has filled sufficiently it overbalances and swings the slot G across the whole width of the sluice, the sample running through to the bucket. It is now in position to be moved again in the opposite direction when the opposite compartment has filled enough to overbalance. By regulating the flow of water at D the time interval between samples may be regulated.

Battery samples taken from the lip of the feeder are necessarily approximate, for unless there is preliminary crushing to at least half an inch product, any one sample may represent rich fines and poor coarse material or *vice versa*. Here again unless done automatically by a sampler that cuts out a certain proportion of the stream of ore the sampling must be left to the caprice of the millman. When amalgamation is not conducted inside the battery a good sample may be had from the lips of the mortar, but when mercury is fed inside the battery the sampling shows only what metal issues from the mortar.

It is a good plan to post the assays of tailing samples every day in the mill so that the millman may know just what he is doing. He is supposed to carry out a system that by careful experimenting has been found to give the best results, but his daily work often causes him to make simple variations from the routine work and it is well for him to know whether these variations have caused a loss or a gain as shown by his tailing assays, otherwise he is working in the dark. While the mine manager must and should be a good millman, he only spends a small portion of his time in the mill and should credit his millmen with an intelligent interest in the work of the mill and the publicity of the mill tailing assays is the simplest method of promoting that interest, without which the results are liable to be inferior.

Battery Feeding

Automatic feeders are so generally known and used that this subject will not occupy us long. It is claimed that by their use a saving of at least 25 per cent. has been gained in the crushing capacity of a mill over that possible with hand feeding.

There are many types of automatic feeders and a careful selection of that best suited to the character of ore to be crushed is a subject of great importance when designing a milling plant.

The main point to be observed is to feed an even product to the mill both as regards the value of the ore and the size to which it has been crushed before going to the feeders. When the grade of ore is constant the millman can amalgamate more intelligently and when the size of the pieces are uniform the feeders work better and require little or no adjustment. Where the single unit stamps are used the feeder is necessarily small and may easily choke if the ore is coarse or wet. As a rule the revolving plate type of feeder is preferable for both the single and five stamp unit batteries, particularly when the ore comes to the mill in a wet condition. The mechanism by which the plate is made to revolve will be discussed later.

The quantity and the quality of the work done in the mill depends to a great extent upon the feeding of the stamps. As a rule the closer the feed, or the nearer the shoe and die come together the greater the amount of ore crushed. They should never "pound," that is, come directly in contact, steel on steel, for that causes vibrations in the stem that ultimately cause a fracture. In an amalgamating battery too much feed will scour the chuck-block plate and may cause a shoe or boss to come off in any battery. Cases where there are not any new stems at hand and where it would be difficult to have one tapered to fit the bosses it is excusable to feed a trifle high to protect the stem against a possible variation of the feed, but high feeding is always accompanied with low tonnage. Ore that is fed to the batteries in a very fine state requires a higher rate of feed than an average crushed ore for the stamp will go further toward the die, having nothing to stop its fall. This accounts for the difficulty of putting tailings through a battery when they have already passed a thirty or forty mesh screen, but if these sands are fed heavy or better mixed with some coarse ore they mill readily enough.

CRUSHING CAPACITY

The number of tons of ore crushed by a stamp mill is broadly speaking proportional to the amount of energy developed by the fall of the stamps, providing its discharging capacity is equal to or greater than its crushing capacity. We have heard of one case at least where it was claimed the maximum capacity was obtained at ninety-six drops per minute, any increase of speed having no effect on the number of tons crushed, but whether this was due to the screening capacity being limited was not stated. It was stated that the higher speed did not allow sufficient time for the ore to settle on the die between the drops of the stamps. Using a single unit battery where the screening capacity was greatly in excess of the crushing capacity the amount crushed per minute increased in proportion to the number of drops, up to 115 per minute, the limit at which the mill could run with a 5 1/2 in. drop.

When the stamp mill is used as an amalgamating machine as well as a crushing machine, the crushing capacity should not be greater than its amalgamating ability, except when it is experimentally found advisable to crush a large tonnage with less saving than a smaller tonnage with a higher saving. This will depend to a great extent upon the grade of the ore and the amount of such ore available in the mine.

Analyzing the crushing capacity of a mill more closely we find that the number of drops and the weight of the stamp are not the only factors. The amount of ore crushed by a stamp mill depends upon the weight of the falling stamp, the number of drops per minute, the height from which the stamps drop, the area and size of the screen openings, the height of discharge, which is the vertical distance between the top of the die and the bottom of the screen opening, the number and character of the splashes produced by the fall of the stamp, the shape and inside area of the mortar, the condition of the wearing surfaces of both shoe and die and the amount and method of feeding the ore and water to the battery.

To increase the amount of ore crushed per stamp one may increase the weight of the stamp or the height of the drop, but the greater loss due to breakages and the less speed at which the stamp can fall with the higher drop are disadvantages. If the ore is of such a character that it will crush with a 6 in. drop nothing can be gained by having a higher drop with a consequent decrease in the number of drops. It is better to increase the weight.

Henry Louis says that the power required to lift a stamp varies directly as the height, while the effective force varies as the square root of the height. In other words the best results are obtained by increasing the weight and reducing the height of the drop. This is being done in the latest constructed mills in South Africa and America. While the increase in California has seldom gone beyond 1250 lb., that in South Africa has gone to 2000 lb. Doubts have been raised whether this weight will be found economical. There is no economy feeding these heavy stamps with material crushed as fine as that fed to the usual 1000 lb. stamp and it is doubtful whether this heavy stamp can do this heavy work as well as the rockbreaker.

Mr. W. A. Caldecott, who is an authority on stamp milling in South Africa, claims for the heavy stamps weighing from 1500 to 1630 lb.

(1) Reduction of the initial capital expenditure in erecting, say, 200 stamps at 1750 lb. with accessories, in place of 280 stamps at 1250 lb. each; (2) reduction in size of mill building, almost proportionate to the less number of stamps; (3) 30 per cent. less shafting, belts and other moving parts to maintain; (4) 30 per cent. less labor required for dressing plates, lubricating moving parts, changing screens, and other work incidental to milling operations.

The difference between the weight of the California stamp and the South African is a natural consequence of the work required by the mill. The former mill is required to crush the ore to a thirty or forty mesh, while the latter crush to ten or twenty mesh. This difference in the ultimate size of the crushed ore accounts for the difference in weight and the duty per day. The heavy stamps are not economical for fine grinding nor the ligher stamps for coarse crushing.

Few of our California mills have a cyanide plant to follow the stamps so it has been found advisable to keep the weight of the stamps within moderate limits, and not crush faster than the ore will amalgamate. We have not yet arrived at the point where our stamps are simply crushing machines and where we can afford to allow our tailings to contain an appreciable amount of gold. In South Africa it is different, for that portion of the gold escaping amalgamation is taken care of in the cyanide annex and the object is to crush as fast and as economical as possible. The work of crushing from twenty mesh to finer meshes can be done much better in a re-grinding machine than in the heavy stamp batteries.

The Colorado type of mill differs materially from the California due to the difference in the physical and perhaps chemical condition of the gold in the ore. In the former type the ore is kept longer in the motor where amalgamation largely takes place. For this purpose the weight of the stamp is about 750 lb., drop about 16 in., speed thirty drops per minute and the discharge about 13 in. In California the weight is from 850 lb. in the older mills to 1050 in those of later date, speed 95 to 105 per minute, drop 6 to 8 in. and the discharge from 4 to 6 in. The latest mill in South Africa is one of 2000 lb. drop about 7 in., speed 98 to 100 per minute and the discharge from 3 to 4 in.

The heaviest successful stamps used in the United States are the 1600 lb. stamps erected at the Vulture Mines, Arizona made, by D. D. Damerest Co., of San Francisco, a description of which appears in the chapter on construction. The next heaviest stamp is the Nissen 1500 lb. stamp of the Boston Con. Copper Co., in Utah. This has a duty of 9 tons per stamp per day through a twenty mesh screen. It is simply a crushing battery and had nothing to do with amalgamating or the use of cyanide.

If we consider the force of the blow instead of the weight of the stamps, the steam stamps used at the Lake Superior Copper Mines are probably the heaviest known. They drop 24 in. at 103 a minute, and crush from 500 to 600 tons through a five-eighth mesh screen. The falling weight is equal to about 5000 lb. We are not considering other than gravity stamps so these will require no further mention for gold and silver ores are totally different than the Lake Superior copper ores.

It is difficult to gather data of the relative amount of ore crushed by different weights of stamps on the same ore, but here is an item furnished by Mr. E. Girault at the San Rafael mill, Mexico.

850-lb. stamps, 7 3/4-in. drop, 104 per min. through 10-mesh 3 to 3 1/2 tons.
1250-lb. stamps, 6 3/4-in. drop, 104 per min. through 10-mesh 6 to 6 1/2 tons.
1250-lb. stamps, 6 3/4-in. drop, 104 per min. through 12-mesh 5 to 5 1/2 tons.

Discharge

The "height of discharge" or the vertical distance between the top of the die and lower edge of screen opening is a most important factor in gold milling. It determines the length of time a given particle of ore is held in the mortar subject to amalgamating influences or to the disintegrating power of the stamps, and therefore is as important a factor on the ultimate size of the issuing pulp as the mesh of the screen. The amount of ore going through a screen, varies inversely as the square of the size of the openings and so far as I can frame a mathematical law by actually varying the discharge with the other factors constant nearly inversely as the square root of the height of discharge, although this latter factor is influenced by the interior area and shape of the mortar. Prof. Richards mentions a case when the discharge was 6 in., the output was 1.8 tons and when 10 in. 1.3 tons, which nearly conforms to the law above mentioned.

The following tests were made at the Southern Cross Mill, Butte County, California, to determine the correct height of discharge and the influence of different character of screens.

Each test extended over a period of six days. The amount of ore crushed per stamp per hour was determined twice daily by noting the time required to fill a 20 gal. tub, allowing to settle, decanting the clear water, drying and weighing. The figures are the average of twelve such tests, at each particular discharge.

TESTS

No.	Screen	Size of opening inches	Discharge inches	Tonnage per stamp per lb.-hour.
1	No. 6 wire..........	0.027	6	300
2	No. 4 blued steel.....	0.035	6	270
3	No. 6 wire..........	0.027	5 1/2	320
4	No. 0 tin...........	0.024	5 1/2	131
5	No. 4 blued steel.....	0.035	5 1/2	280
6	No. 6 wire..........	0.027	4 1/2	340
7	No. 0 tin	0.024	4 1/2	146

These tests very nearly conform to the law above enunciated. The usefulness of knowing such a law is apparent when it is necessary to figure on an increase of crushing capacity. Suppose that I am crushing 20 tons of ore a day and I strike a body of low grade material which may pay if put through the mill fast enough. I have two courses open, either to use a screen with larger openings or lower the discharge. Suppose a coarser screen is out of the question and I wish to figure the possible tonnage with a lower discharge. The square root of 6 is 2.449, of 4 is 2. The number of tons per day—20 multiplied by 2.449 is 48.98, divided by 2 is 24.5. Therefore by lowering the discharge from 6 to 4 in. I increase the capacity of the mill by 4.5 tons.

For a given ore there is a point at which a certain height of discharge will be the most efficient and as near as possible to this point it should be kept as the dies wear down. This can be done by chuck-blocks of different sizes, strips of wood nailed on the screen frame, having top and bottom of screen frame of different widths and reversing, by raising the chuck-blocks by means of strips of wood between it and the mortar or by putting false dies under those that have worn down. The oldest mills we know of, those described by Agricola had no screens, the height of discharge determining the size to which the ore was crushed. "The grates (screens) were entirely dispensed with, and the pulverized mineral was projected over a small wooden shutter, fixed in lieu of a grate. This shutter was raised or lowered in the grate holes, according as the ore was required in a rough, fine or a coarse state."

With the fast crushing single unit mills the discharge should be kept as low as possible, the determining factor being the wear of the screens, for the object is to crush as much ore as possible depending upon the screens to size the pulp, and as amalgamation only takes place on the outside, the ore is not kept in the mortar any longer than necessary.

It may often be noticed when a soft or argellaceous ore is being crushed that the screens clog easily and the mortar soon fills up. The fault may be in the screen or it may be that the discharge is too high and by simply lowering this a couple of inches a great improvement is noticed. Where a chuck-block with an inside plate is used the discharge must be kept sufficiently high to prevent scouring of the amalgam, but not high enough to unnecessarily retard the crushing. It may be found beneficial to use a coarser screen with a higher discharge, but this must be determined by experiment for there is no mathematical law or set of rules that one can follow.

Screens

The material of which screens are made as also the mesh most desirable should be questions of experiment, for what would be

satisfactory with one ore or a certain shape of mortar may be
totally unsuited for another ore or mortar. Some use brass
wire; others Russia iron, angle or straight slot or perforated
steel; others burnt tin or blued steel, while again others use
steel or brass wire screens. The blued steel needle perforated

Fig. 6.—Diagonal slot screen.

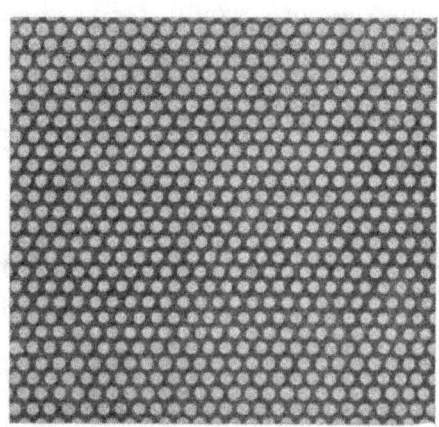

Fig. 7.—Needle punched screen.

Fig. 8.—Horizontal slot screen.

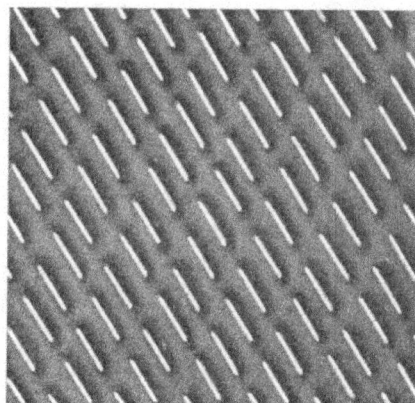

Fig. 9.—Diagonal burred slot screw.

screens are the same thickness as the burnt tin and give a
good discharge even with the fine meshes, but last only from
two to four days. Considering the free delivery and low cost
of these screens they are very economical for fine meshes while
the steel wire screens are certainly very good for quartz ores

but for clay or talc ores they are apt to clog. For this latter class of ore the Russia iron or steel slotted screens are better. An ore that contains partially decomposed copper sulphides may quickly destroy a steel or iron screen, while a brass wire would wear well. This emphasizes the fact that each individual ore must be treated on its own merits, and what would do well on one ore would be a disastrous failure on another. An instance will suffice to clinch this truism. A certain large mill was using diagonal slot steel screens and as an experiment a horizontal "long ton" steel wire screen was tried. Within an hour every mortar in the mill was filled to the top, the stamps pounding iron. The crushed ore would not go through the screens. The old screens were quickly replaced. Another experiment was made, this time using a "long ton" screen of the same make, but with the opening in the wire vertical instead of horizontal. This answered fairly well, but the old diagonal slot screens proved to be the most satisfactory both as regards to the tonnage and the wear.

The following tests on a pure quartz ore prove conclusively the superiority of the steel wire screen for this particular ore:

TESTS

No.	Screen	Discharge inches	Assay of tailing	Tonnage per stamp per lb.-hour
1	No. 6 steel wire......	6	$0.90	300
2	No. 4 blued steel.....	6	1.20	270
3	No. 6 steel wire......	5 1/2	trace	320
4	No. 0 tin............	5 1/2	0.90	131
5	No. 4 blued steel.....	5 1/2	trace	280
6	No. 6 steel wire......	4 1/2	1.45	340
7	No. 0 tin............	4 1/2	1.85	146

In the case above cited the steel wire screens not only gave the highest tonnage, but the lowest tailings.

Slotted iron screens generally give a coarser product than the number would indicate, while the needle point perforated give a finer pulp. The former soon wear enough to open the slot appreciably, while the latter soon clog with pieces of crushed ore. In the first case the thicker the metal the less the initial wear and in the latter case the thinner the metal the less clogging.

Wire Screens

Gauge.—The "Washburn and Moen Gauge" is the standard for iron, steel and tinned wire cloth and "Old English Gauge" for brass, copper and bronze wire cloth.

Mesh.—The word "Mesh" in wire cloth means the number of openings per lineal inch, measured from center of wire, and does not indicate the size of the openings unless the size of the wire is known. For example a thirty-mesh steel wire cloth may be made of wires between No. 27 and 36 giving size of openings between .0163 and .0243 of an inch, or a twenty-four mesh screen with No. 24 wire will give about the same size opening as a thirty-five mesh cloth with No. 34 wire.

Heavy steel cloth for mills is usually made of the following size wire unless otherwise ordered:

Number meshes per inch	Number wire	Size of openings	Compared with H. & H. perforated metal screens.
....	....	.058	No. 1 slot
....	....	.049	No. 2 slot
....	....	.042	No. 3 slot
18	26	.037–.035	No. 4 slot
20	26	.032	No. 5 slot
24	32	.028	No. 6 slot
24	28	.025	No. 7 slot
26	28	.022	No. 8 slot
28	29	.020	No. 9 slot
30	29	.018	No. 10 slot
35	31	.015	No. 11 slot
40	32	.012	No. 12 slot
50	35	.0105	....
60	36	.0076	....
70	38	.0062	....
80	40	.0055	....

TABLE SHOWING DIFFERENCE BETWEEN WIRE GAUGES

No.	Old English	Washburn & Moen	Approximate size	Birmingham	American B. & S
0000	.454	.393	....	.454	.460
000	.425	.362	....	.425	.40964
00	.380	.331	....	.380	.36480
0.34	.340	.307	....	.340	.32495
1	.300	.283	9/16	.300	.28930
2	.284	.263	....	.284	.25763
3	.259	.244	1/4	.259	.22942
4	.238	.225	7/32	.238	.20431
5	.220	.207	13/64	.220	.18194
6	.203	.192	....	.203	.16202
7	.180	.177	....	.180	.14428
8	.165	.162	5/32	.165	.12849
9	.148	.148	....	.148	.11443
10	.134	.135	....	.134	.10189
11	.120	.120	1/8	.120	.09074
12	.109	.105	....	.109	.08081
13	.095	.092	3/32	.095	.07196
14	.083	.080	....	.083	.06408
15	.072	.072	1/16	.072	.05707
16	.065	.063	....	.065	.05082
17	.058	.054	....	.058	.04525
18	.049	.047	3/64	.049	.04030
19	.040	.041	....	.042	.03589
20	.035	.035	....	.035	.03589
21	.0315	.032	....	.032	.02846
22	.0295	.028	....	.028	.025347
23	.027	.025	....	.025	.022571
24	.025	.023	....	.028	.0201
25	.023	.020	....	.020	.0179
26	.0205	.018	....	.018	.01549
27	.01875	.017	....	.016	.014195
28	.0165	.016	....	.014	.012641
29	.0155	.015	....	.013	.011257
30	.1375	.014	....	.012	.010025
31	.1225	.0135	....	.010	.008928
32	.01125	.013	....	.009	.00795
33	.01025	.011	....	.008	.00708
34	.0095	.010	....	.007	.0063
35	.009	.0095	....	.005	.00561
36	.0075	.009	....	.004	.005
37	.0065	.0085	....	....	.00445
38	.00575	.008	....	....	.003965
39	.005	.0075	....	....	.003531
40	.0045	.007	....	....	.003144

The following comparison sizes of English and American Stock Sizes of Wire Cloth is taken from T. J. Hoover:

COMPARISON OF ENGLISH AND AMERICAN STOCK SIZES OF SCREEN CLOTH. T. J. HOOVER

Cube root series	Potter's stock, London			Tyler's stock, Cleveland, Ohio		
Aperture, inches	Aperture, inches	Size of wire, inches	Meshes, per inch	Aperture, inches	Size of wire, inches	Meshes, per inch.
0.1250	0.1270	0.04	6	0.1270	0.04	6
0.0992	0.0990	0.026	8	0.1000	0.025	8
0.0787	0.0790	0.021	10	0.0770	0.023	10
0.0625	0.0630	0.020	12	0.0630	0.0205	12
0.0496	0.0490	0.022	14	0.0488	0.01375	16
0.0394	0.0407	0.0148	18	0.0397	0.01025	20
0.0313	0.0318	0.0316	22	0.0314	0.01025	24
0.0248	0.0241	0.0116	28	0.0247	0.01375	26
0.0197	0.0193	0.0092	35	0.0195	0.01375	30
0.0156	0.0156	0.0090	40	0.0155	0.0095	40
0.0124	0.0124	0.0076	50	0.0125	0.0075	50
0.0098	0.0099	0.0068	60	0.0098	0.01025	50
0.0078	0.0075	0.0050	80	0.0077	0.0065	70
0.0062	0.0063	0.0048	90	0.0061	0.0050	90
0.0049	0.0047	0.0036	120	0.0050	0.0040	110
0.0039	0.0035	0.0020	180	0.0039	0.0032	140
0.0031	0.0034	0.0016	200	0.0031	0.0024	180
0.0025	0.0028	0.0014	250	0.0030	0.002	200

The cost per ton for screens varies in different mills, an average being about 1 1/4 cent. The particular screen to use is that which will give the least value in the tailings with the maximum output if there is no after treatment of tailings. There may be cases where it is preferable to increase the output at the expense of the saving capacity, but it will be wiser to increase the number of stamps. If the tailings are to be cyanided in any case it may be economical to make a finer or a coarser product according to the susceptibility to cyanide treatment or more economical to grind very coarse and regrind in some other machine to a finer mesh. Often a change to a coarser mesh may be beneficial in every respect. A particular case may be cited where the change from a fifty mesh to a thirty mesh screen not only gave a more suitable cyanide product, but decreased the residual values going to the cyanide plant about one-third. The mill had been sliming both the quartz and the gold and the cyanide plant was built for sand treatment, not slimes, and the extraction by cyanide on the thirty mesh product was as good as when ground finer. This is not the usual result of fine grinding I know but this was an, exceptional case.

Screen Frame

The screen frame is usually inclined outward at an angle of 10 to 12 degrees to facilitate the discharge of pulp through the screen. It also prevents undue wear of the screen by causing the particles of ore forced from under the stamp to hit the screen at a less. angle than if the screen was vertical. The five stamp batteries have from 3 1/2 to 4 sq. ft. of screen surface, while the single unit batteries have about the same amount or five times the screen surface per stamp. As the lower 5 or 6 in. does most of the screening the frame may be made to receive a screen a foot wide. This will allow of the screen being reversed when worn on one side and also leave a space between it and the top of the mortar for a head board, which may be held in by a different wedge than that which holds the screen. This head board can then be lifted out without disturbing the screen.

The excessive wear of the screens in the single unit batteries is caused by the vibration of the screens due to the suction caused by the stamp. Screens that have had very little wear from contact with the ore particles are seen to crack near the frame from this bending movement caused by the pulsations within the battery. It is doubtful if this can be altogether overcome. It appears to me that a greater inclination of the screen from the vertical might help this by enlarging the mortar above the level of the water and so give space for the air to expand and contract and at the same time, where in the case of the single unit mills the die is so near the screen, the angle of splash from the die to the screen being less might not be so destructive.

The screen frame should be made with at least one vertical rib and better two. This will decrease to a limited extent the screening space, but would save breakages due to the vibrations.

Scroll or the Order of Drop

Much confusion exists in the minds of millmen as to the relative merits of different orders of drop. This would not be so if the stamps were numbered in the order in which they drop from both right to left and left to right. It would then be seen that for a given order of drop in one direction there is a different order in the other direction, so that those advocating 1–4–2–5–3 are also advocating 1–3–5–2–4 for they are identical and the only two orders of drop where no two stamps drop in succession. This is generally known as the Homestake drop and requires the end stamps to have a slightly greater drop than the others. Stamps are usually numbered from the driving pulley and when no other method is mentioned this is understood. The scroll, 1–5–2–4–3, or 1–4–2–3–5 is known as the California drop, where two stamps drop in succession. The object is to obtain a regular wave in the mortar with every stamp doing the same quality of work. In roomy mortars the end stamps usually require a little higher drop no matter what the order of drop. The California drop while not crushing as much as the Homestake gives a more regular wave and consequently better inside amalgamation and is not so sensitive to changes of speed.

There is a secondary drop as described by Mr. R. Bell, obtained by omitting every other stamp in the sequence. In the 1–3–5–2–4 sequence the secondary drop is 5–4–3–2–1 or a tertiary drop by omitting every two stamps showing 1–2–3–4–5. From an academic point of view and as a mathematical problem these intermediate drops may be interesting, but in practice neither the height of the drop nor the weights of the stamps are all the same and the millman must set his tappets not by measuring the distance between shoe and die in the mortar, but on the cam-shaft floor with no other guide but his sense of touch and his own experience. Be this as it may my experience with the sequence 1–4–2–3–5 has been satisfactory. Assuming for argument that the secondary drop is important, in this sequence it would be 1–2–5–4–3, which would suggest that the tendency of No. 5 to choke would be as noticeable as in the sequence 1–4–2–5–3. The speed at which the stamps drop has an influence on the sequence, for this speed will determine the number of splashes and the time that the next dropping stamp meets the wave produced by some previous splash. In practical work it is often noticed how poorly the battery works when the speed has been altered and any permanent variation from the normal speed may require a resetting of the tappets or an alteration in the feed.

When ten stamps are attached to one cam-shaft the stamps in the second mortar follow the stamps in the first and with the same order of drop, thus, with the scroll 1–5–2–3–4, the ten drop in this order, 1–6–5–10–2–7–4–9–3–8.   To equalize the thrust on the cam-shaft the cams of the two sets are arranged on different sides of the stems so while one set throws the cam-shaft to the left, the other throws it to the right.   Where five stamps are on a cam-shaft the end thrust is taken up by the collar on the end.

Width of Mortar at Discharge

The width of the mortar at discharge level varies from 11 in. to 18 in., these being the extremes for rapid crushing and amalgamating batteries.   The more narrow the mortar the quicker will the crushed ore be expelled from the mortar and the less chance there is of using inside amalgamating plates for the scouring effect of the splashes becomes intense in an 11 or 12 in. mortar while in an 18 in. mortar the splashes cause only a wash over the chuck-block plates.   Single unit batteries being made for fast crushing are always made narrow.

The standard amalgamating mortars are from 17 to 18 in. wide at a 6 in. discharge.

Wear of Shoes and Dies

The uneven wear of shoes and dies causes a falling off in the amount of ore crushed.   This needs little explanation for it is evident that the more true the wearing surfaces the more ore will be in the position where it will have the full effect of the falling shoe.

Mr. Urquhart expresses the opinion that the uneven or irregular wear of the shoes and dies in stamp batteries may be prevented by the use of water directly upon the dies.   The wear of the die he states, "is in accordance with the order of drop, providing that the weight and height of the drop are the same. Thus with an order of drop 1–3–5–2–4 dies No. 1 and 2 wear least and hollow most, Nos. 4 and 5 wear most and hollow least because Nos. 1 and 2 always have more ore upon them than Nos. 4 and 5.   With the order of drop 1–5–2–4–3, Nos. 4 and 3 wear least and hollow most, Nos. 1 and 5 wear most and hollow least for the same reason."

"With the same order of drop when the diagonal water supply is used, all jets being the same, the results so far as wear is concerned are practically the same, but the dies are less inclined to hollow because the water playing upon them washes off the crushed particles and leaves clean ore upon clean dies for the next blow."

In his opinion the cause of the "caving" is the fact that after the stamp has struck its blow, the crushed particles on the dies are gradually ground in, new particles being added with each blow of the stamp. The remedy is the use of sufficient water distributed in proportion to the requirements necessary to clean each individual die. When the ore banks most the larger quantity of water is used. By using too much water a slightly rounded effect is produced on the die, exactly the reverse of that when the water is used too sparingly. If it is desired to take the "cup" out of a die, it is advisable to feed fine and use plenty of water.

There is no doubt that the water feed, that is the method of feeding water in the battery, will have an appreciable effect upon the wear of the dies, but it is seldom advisable to alter the amount of water from that necessary to clear the plates of all the sand, except where the plates are of secondary importance and the water supply abundant.

My experience has been that the uneven wear of shoe and die is caused to a great extent by the hardness of the material of which they are made. Chrome steel shoes and chrome steel dies wear unevenly at the expense of the shoe. It is so much more convenient to replace a die than a shoe that it is better to let a die wear in preference to a shoe. Therefore I use a die of a material not quite as hard as the shoe. This not only gives the die most of the wear, but I find it has the effect of producing a more even wear on both shoe and die. Where both shoe and die were of the same hardness of chrome steel I have seen the dies cup with hollows 6 in. deep, almost the full depth of the new die being hollowed out.

To keep the weight of a stamp up to a given standard as the shoes wear down, compensating weights may be used, clamped to the stem above or below the tappet, preferably the former. These weights may be made of two pieces and clamped to the stem as the feed-collar is clamped and may weigh about 50 lb. When the shoe has worn down about 50 lb. one of these weights is fastened to the stem. This idea I believe comes from South Africa. I have never seen or heard of them in use in the United States.

Area of Shoe and Die

No comprehensive tests have yet been carried out to determine the ratio of the shoe and die superficial area to the weight of the stamp. The mesh to which the ore is crushed must be considered in any such test, and so it becomes rather a complicated problem. The heavier stamps require larger stems and tappets and consequently the area of the die has been increased slightly, but it is doubtful whether capacity can be gained by any great increase in the area of the die, for the cushion of ore is increased, while the pressure per square inch of surface is decreased. To attain a high duty the blow of the dropping stamp should not be cushioned more than necessary for it is obvious that the less ore on the die up to the point when pounding begins will supply sufficient ore for crushing.

While the usual 1050 lb. batteries have a shoe diameter of 9 in. the 2000 lb. stamps have 9 1/4 in. As the die is usually made a quarter of an inch to a half larger than the shoe for an increase of weight of nearly 100 per cent. the increase in the area of the die surface has increased about 3 per cent.

Water

The amount of water used for milling purposes varies between very wide limits, according to the nature of the ore and the after treatment of the tailings. For crushing alone the amount will vary from six to ten times the weight of the ore stamped. The California mill of 850 to 1050 lb. uses from 5 to 7 gal. per stamp per minute for all purposes except power. Where water is scarce and a tailing pond is used to settle the pulp and the clear water is pumped back for reuse, about 1 gal. per stamp per minute will suffice, depending upon the character of the ore not being of a soft slimy nature. When the machinery is run by distillate engines and the tailings are drained in tanks from 90 to 100 gal. per ton of ore crushed has been found sufficient.

The water used in amalgamating mills is that just sufficient to take the pulp off the plates and should be fed within the mortar. The old style of having a stream of water on the outside plates has been discarded in favor of using all the water inside the mortar.

Salt water may be used in the batteries with no detrimental

effect and sometimes where oxidized copper compounds occur in the ore it may be an advantage to add salt to eliminate these compounds from the cyanide tanks. The bullion from the amalgam will be baser by the amount of copper precipitated on the plates.

Many mills use a circulating solution of cyanide of potash, that is the ore is crushed in cyanide solution. Usually the amount of such solution will be from 6 to 8 tons of solution per ton of ore.

Power Required to Drive Mills

By dynamometer measurements Von Reytt obtained a ratio of theoretical horse-power used in overcoming gravity plus friction of 1.1:1.27. Henry Louis computed this value at 1:1.202. Taking Von Reytt's figure, the actual horse-power to operate twenty stamps run under the most perfect conditions would consume 42 1/4 h. p.

$$\text{Horse-power} \frac{\text{Weight of stamp} \times \text{drop in feet} \times \text{drop per minute.}}{33,000}$$

1000. lb. $\times$ 7/12 ft. $\times$ 106   1.8735 h. p. $\times$ 20   37.47 h. p.   37.47 $\times$ 1.127   42.22 h. p.

Sometimes the horse-power is reckoned on the number of tons crushed as the figure given for the San Rafael mill of 1.68 h. p. per day per ton of ore crushed for stamps and tube mills.

Cost of Stamp Milling

The cost of crushing and amalgamating a ton of ore varies so greatly with the conditions, such as the character of the ore, the price of labor, cost of power and supplies, the amortization of the capital expenditure and the interest on the capital invested that without knowing these conditions there is no way of arriving at a fair estimate. The cost of crushing with a five-stamp mill may vary from one dollar to two dollars per ton and with a hundred stamps from 30 cents to 60 cents per ton.

The usual estimates found in text books are not all clear as to what is included in this cost. The cost of superintendence is seldom included and the gradual return of the amount invested is never reckoned at all.

Many failures of mining ventures are caused by inattention to the fact that the capital invested as well as a fair interest must be returned before the mine is exhausted. If I erect a mill on a property that has blocked out 20,000 tons of ore and a possibility of another 20,000 tons the cost of milling should include the cost of putting up that mill for the 40,000 tons of ore. Say the mill cost $20,000, then besides labor, power supplies and superintendence there must be added 50 cents a ton for amortization of the outlay. If this factor was considered a little more there would be fewer mills without mines and more mines with mills.

CHAPTER III

PRACTICAL WORKING OF THE STAMP MILL

The ore coming to a well designed mill may have been previously broken to 1 or 2 in. cubes or if it has not been previously crushed·it is dumped upon a grizzly (grate) to separate the coarse rock from that portion requiring no further crushing before going to the stamps. This coarse portion is fed to a rock breaker and both this and that previously gone through the grizzly are mixed in the storage bin below.

In a small mill it is economy to have the rockbreaker of a greater capacity than that actually needed, for it not only allows the breaker man time for other work, but it is on hand if the capacity of the mill is increased, either by the addition of more stamps or the addition of another type of regrinding machine. If the breaker be large enough all the ore for the twenty-four run may be broken during the day shift. The argument against this is the increased size of engine needed, but the difference in first cost will soon be more than compensated by the gain.

A five stamp mill or mills of a few small units can only be considered as experimental plants, very few of them pay to run very long and the question of enlargement must be considered sooner or later, so a little extra expense at the start will save enough to warrant the conversion.

Where the mill is of small capacity I have put the rockbreaker on the same level as the battery with an elevator to the ore bin. This saves the labor of a rockbreaker man, for the millman has plenty of time to feed the breaker. The cost of elevating the broken ore is less than the labor of an extra man.

We see in reports that a mill has run a certain per cent. of the time, the stoppages being due to a variety of causes some of which could be reduced by careful attention to keeping the mill always in repair, and when repairs are necessary to have them done well. The time expended in doing a good job will be repaid in the gain in running time. If we allow thirty minutes a day for two brushings of the plates and one and one-half hours per battery once a month for cleaning up this will consume for a 100-stamp mill about 1650 stamp hours per month, or say 2 1/2 per cent. of the time; all other losses are due to necessary repairs.

A mill to attain its maximum capacity should be at work all the time, except, of course, such time as is actually necessary for keeping the plates in condition for catching the precious metals, for the regular cleanup, and, for changing shoes and dies about once in every six months. This last item does not consume more than fifteen minutes per stamp twice a year and so is hardly appreciable. The die need only be changed on cleanup days and it takes no more extra time than setting the tappet, which must be done anyway for the dies had better be turned half round. This will help to equalize the wear.

Much time is often lost by not having duplicate parts handy. This is poor management and inexcusable, but cannot always be blamed on the superintendent, for he is often denied the privilege of purchasing what he deems necessary.

The time necessary for brushing up every day varies in different mills. Some plates foul easily and may require dressing every four hours, others keep bright for twelve hours. The rule is to dress the plates when the surface shows the need of cleaning the amalgam. This again is a question for experiment on the ore under treatment.

Arrangement of the Factors

The five factors, weight of stamp, height of drop, number of drops per minute, character of screen and height of discharge should be so correlated as to produce the best results on the ore under treatment. The weight of the stamp as the shoe wears down is changing all the time unless compensating weights are used, but the other four factors can be regulated at will. Having had no previous experience with a battery it were best to keep the drop below **7** in., for with a heavy stamp at least the higher the drop the more liability of a broken stem. With new shoes it is advisable to start with a drop not over 5 1/2 in. This will give a chance for the shoe to conform to the shape of the die and will obviate any excessive "pounding" at the start.

As the maximum height of the drop is fixed by the shape of the cam this should have been determined before the mill was ordered.

The height of discharge is perhaps the most difficult factor to determine. There is no rule to follow and each ore must be experimented upon to determine this point. If the ore must be amalgamated inside the battery we know at once that the discharge must be over 5 in. and not so high that the tonnage going through the screen will be too small. If the battery is a crushing machine simply with no amalgamation inside the mortar then the discharge must be kept low to prevent sliming and to keep up the tonnage, but not so low that there is undue wear on the screens. For this latter class of mill the discharge should not be less than 2 in. nor over 4 in.

Broken Cam-shafts

Cam-shafts sometimes crystallize or change structure so that they break. This is due to vibrations caused by inattention to keeping the shaft in alignment, but even if this is carefully attended to a break may occur, but no doubt if traced back far enough it will be found to be the fault of some neglectful millman. Every mill should have a spare shaft preferably ready fitted with cams and should one break it may be replaced with the new one or the old one stripped and the cams put on the new shaft, hoisted into the cam-boxes and the bull-wheel put on. Much time is lost in taking off the old bull-wheel which except in the later built mills are invariably fastened on with a key. Now we have the bull-wheel put on the same as the cams with a curved brass gib. It is then no trick to take off a bull-wheel and replace it on another shaft. I have seen a keyed wheel that required the services of four men a full day to disconnect it from the shaft. It was the same with the old keyed cams, they either were keyed so fast that it was a day's job to take one off or they were never keyed firm enough to last a shift.

Cam-shaft Collars

Cam-shaft collars loosen by vibration of the cam-shaft; the remedy is to line up the shaft, thus reducing the vibrations to a minimum. It will be well to have a split collar handy with two set-screws to replace a collar that continually slips, especially if it must replace one next to the bull-wheel. These collars are for the purpose of keeping the cams near but not scraping against the stems and it is most important that they should always be kept at their proper place.

Changing Stems

Stems break from change of structure, called crystallization by some, and a succession of minute cracks by others; at all events it is due to the vibrations caused by the falling stamp and to the lift of the cam. The break seldom occurs but at the two nodes where the stem enters the boss and where it enters the tappet, except of course when stems have been welded when the break is liable to be in the weld, being in this case the weakest point. The uneven wear of shoe and die, the play in the guides causing blows at an angle to the line of drop and the introduction into the mortar of mine or mill tools, may cause a broken stem.

Broken stems should be annealed before being used again to correct any crystallization remaining near the broken end. When a battery is hung up, if the stamp that dropped last is allowed to drop first when starting, the jar due to the shoe striking the die may cause a broken stem. It is therefore best to hang up the feed stem first and drop it first. This throws the ore under the feed stem when hanging up and acts as a cushion when starting.

The quickest way to change a broken stem is to take off the guide caps and turn the stem over if the tappet and stem are not good on the reverse side. If the stem is broken on both ends the tappet must be taken off, put on another stem, fastened at approximately the right height, the stem replaced in the guide and the guide caps bolted on. If there is a repair crew, while this is taking place on the cam-shaft floor the crew on the battery floor have taken off the screen, and rolled the boss-head out of the mortar. The shoe is forced out of the boss-head and may be wedged into the new boss or the boss put in the mortar and the shoe put on afterward. In the latter case when "sticking" the stem in the boss the latter should be protected by placing a piece of board between it and the die or the boss-head may be broken. With the stem in the boss-head the tappet is set at the height which by experience is found to give the proper drop. The shoe is now centered under the boss, wedges put in place and the stamp dropped. The tappet may be set at its proper place at any time after the balance of the stamps are dropping unless it is the feed tappet, when it must be set before starting the battery. With two men on each floor the change may be made conveniently in half an hour and it has been done inside of 15 minutes. If there is no extra boss-head on hand the broken end must be forced out at once with powder or with a wedge; the latter method is preferred because unless one is experienced in the use of powder for this purpose a cracked boss-head is liable to be the result. The battery in this latter case may be hung up for hours according to the luck of the millman or according to his judgment in selecting tools and his manner of using them. If there is an extra boss-head on hand, the broken stem may be forced out at leisure.

Should a stem break on the night shift it may be convenient to wait until the day shift to make the repair, particularly in the larger size mills where the repair crew is only required on the day shift. In this case it is best to take the boss out of the mortar and hang up that stem. Should it be the feed stem the same procedure will answer except that an adjoining stem must be converted into the feed stem. For this purpose an extension arm is kept on hand. This is clamped on the feeder arm, the feed collar changed to the adjoining stem and the four stamps can drop until the day shift makes the repair. In small mills the millman must change the stem by himself, but can keep the battery going with four stamps until everything is ready to replace the stem. There is no economy in keeping the whole mill hung up for hours when it is impossible to judge the length of time required to change the tappet and have a boss ready. Better to take the boss out of the mortar and the stem out of the guide and let the battery run.

Bosses Off

Bosses or boss-heads come off because the drop is either too high for the speed, causing "camming," or too low for the style of cam, causing a series of blows which may force the stem out of the socket of the boss. If there is too much play in the guides so that the bosses strike the stem may come out. It may be that the stem has not been turned to fit the boss accurately, in which case it must be carefully shimmed. Some prefer putting on bosses with iron shims. No doubt this is the only method when there is a poor fit, otherwise canvas is to be preferred. For a drop of 7 in. providing the shoe is already on, the stem is let down until it is half an inch above the boss, the tappet keys driven in, the stem centered over the hole in the boss and the

helper or the amalgamator alone if he has no help drops the
stem with the cam stick under the tappet for the first two or
three drops and then with the cam alone until the stem is well
stuck.   The battery is now ready to close up and start running.
With a little practice using a given height of drop and setting the
tappet so that the stem will be the right distance above the boss
the stamp will be dropping the proper height.   Some millmen
throw off the belt on the bull-wheel either at the friction clutch
or the belt tightener, hammer the stem into the boss and then
set the battery going.   This is a waste of time and would not be
allowed in any mill where the running time is of the least importance.   As a rule stamps are never hung up unless it is absolutely
necessary.   I have stuck many stems in bosses without hanging
up any stems except the one being operated upon.   The same
with tappets, they are all set with the other four stamps dropping.

If the boss is cracked or the socket hole worn through carelessness in not hanging up the battery when the stem broke it
will be necessary to put in a new boss.   If after two or three
trials the boss refuses to stick on the stem examine both the boss
and the stem carefully, for the stem may be burred or the hole
in the boss choked, in either case a cold chisel judiciously used
will be of benefit.

The student may like to know how to be able to know when a
boss is off.   Three conditions may be noticed, the tappet is
camming and the stem jumping up and down light, while all
the other stamps are dropping with their accustomed thud, the
loose stem is dropping only a portion of its height showing that
it is hitting the top of the boss, and the fall of the stem may be as
if it were pulled out of molasses when the stem is entering the
hole in the boss at every drop but not "sticking."   This last
is the most difficult condition to detect.

Shoes Off

Shoes come off from too much play in the guides, too much
ore in the mortar or from the pounding of the shoe on the die.
The fault is generally a personal one.   If the shoe is difficult to
"stick" roughen it with a chisel.   Do not use soft wood wedges;
they will not hold a shoe on the boss and will cause loss of time.
One must judge the thickness of the wedges and the number of
wedges to use by the closeness of the fit.   The shoe should be on

tight but not so tight that it touches the boss. A quarter of an inch clearance will be sufficient. How can one tell if a shoe is off? The stamp will only be dropping a portion of its height or the tappet will be camming. No splash will take place in front of the stamp nor will there be any change in the height of the drop of the stamp if a quantity of ore is pushed into the mortar. These symptoms are the same as when a boss is off but the stem is lighter in the former case and jumps considerably. If for some reason the shoe will not stay on with the hard wood wedges try again with a piece of canvas thrown over the top of the shoe after the wedges are tied on. Some prefer to tack the wedges to a piece of canvas and tie round the shank of the shoe, but the most usual method is to throw a loop of string over the shank and put in the wedges one at a time until all are in place and then fasten tightly.

Dies Out

Dies seldom become misplaced unless they are worn down to near the discarding point and should then be tightly wedged with iron or wooden wedges. The wear on a die is about an inch a month and as it is not economical to throw metal away that may be used they can be used until too thin to wedge in or until a hole wear in the center. The dies wears so differently in mills that what would be a safe limit in one mill would not apply to another. It may be safe to say that a die worn down to under 2 in. at its thinnest part had better be discarded.

Broken Cams

Steel cams seldom break, the only cause being by "camming" due to a loose tappet, a shoe off, a boss of or a "run away" of the engine. If camming occurs the millman should at once hang up that stem and set the tappet; if the fault is in the engine room the engineer will immediately notice it. A broken cam must be immediately replaced. This may mean the cam-shaft must be stripped of all the cams and a new one put on. Self-locking cams are loosened by hammering on the horn in the direction opposite to that when running. It will first be necessary to provide heavy blocks of wood to hold the cam shaft after it has been lifted out of the bearings with the chain-blocks. If the bull-wheel comes off easily it may be best to sling this on a chain

3

and take it off the shaft, especially if the broken cam is near that end.   With the self-locking bull-wheels this is a matter of a few minutes.

Feeding the Ore

Ore is often delivered to the stamps too coarse for economical work.   It is better to feed a finer product, the feeders work better, there is less wear on the screens, shoes, dies and guides and more time for the millman to look after his plates and keep his mill in good repair.   There are limits between which it is economical to work any extra crushing smaller than this size or failure to crush to this size will lessen the capacity of the mill.   As all ores are separate problems no definite size can be recommended but the limits lie between 1/2 and 1 1/2 in. cube.

The millman can judge how his batteries are feeding either by the splash on the screen, by the feel of the stamps, or by the sound of the falling stamps.   To start the feed the best plan is to let the ore run out by hanging up the feeder.   When the stamp begins to pound turn the feed on until the pounding ceases. This is the point at which the stamp is doing its best work, just so the steel of the die does not directly come in contact with the steel of the shoe.   When starting the feed of a battery after it has been cleaned out or when starting with new shoes and dies be sure to rake down plenty of ore to cover the dies.   Stems sometimes break when only pounding for a short time; even three or four blows on a bare die may be enough.

Setting Tappets

Tappets usually slip from the stem having worn too small for the gib to hold.   A tappet that does not fit close to the stem binds only on two surfaces, that at the gib and that on a line opposite to the center of the gib.   Patents have been taken out for tappets with holes bored in other shapes than round in order to give more than two bearing surfaces.   A key-way opposite the gib for the insertion of a key or shim that would fit the stem and give more chance for the gib-keys to hold would be good to have on hand to use on a worn stem.

To set a tappet on a worn stem an iron shim must be driven down between the stem and the tappet opposite to the gib and the keys driven in with a heavy hammer.   This will, no doubt, in a very short time break or otherwise damage the gib when the tappet must be taken off and replaced with another. The battery need not be hung up to change a tappet unless it is the feed-stem tappet or the long wooden guides are used. The best remedy is a new stem, but few mills can afford this luxury. The cause of the stem wearing was probably in the first place due to the guide having too much play, the short iron guides being the most objectionable.

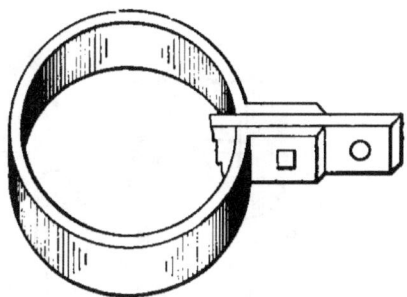

Fig. 10.—Stem lifter (circular).

In one mill I noticed that whenever a tappet was to be set the whole five stamps of this battery were hung up. This practice no doubt came about through the fact that the apparatus for holding the stem, when hoisting on the chain-blocks was a closed ring as shown in Fig. 10. To put this over the stem the millman must climb up on another tappet, and for safety, the whole five stamps were hung up. Even a step ladder was sometimes placed

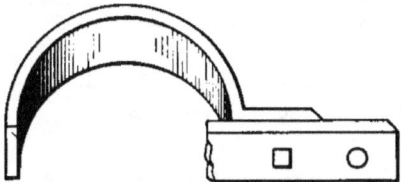

Fig. 11.—Stem lifter (half round).

against the battery post to aid the millman to reach the top of the stem. An open ring made of 2 in. iron or steel, as shown in Fig. 11, would have obviated all this difficulty, and only the stem required to be set, hung up.

Tappet setting may seem inconsequential to some, but it is of vast importance. It may mean time lost and material wasted.

Many millmen lower the tappet by slightly loosening the keys, allowing the cam to shift the tappet on the stem. This is a

quick way and may be justifiable in many cases, but it jams the
gib into the stem and after three or four such settings, it is
nearly impossible to move the tappet without injury. A better
plan is to knock out the keys and let the stem down on the
chain-blocks, or better yet, to clamp a collar the right distance
above the tappet, and let the stem fall until the clamp rests
on the tappet. The keys need not be taken out, only loosened
and   no   harm   is   done   to   the   stem,   gib   or   tappet.
The clamp may be made with an offset to turn a clamp screw as
shown in Fig. 12.

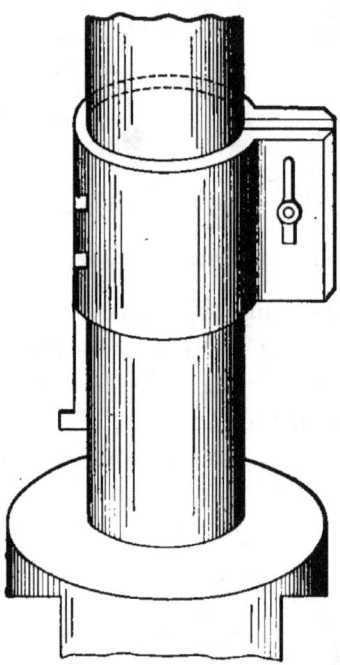

Fig. 12.—Stem clamp.

As a rule it will not pay to "monkey" with tappets that will
not stick on the stem, but it is not every manager that provides
proper material for economical work so that for reasons unknown
to the millman he must use what is put before him. A tappet
may have a perfectly good gib but the stem not fitting close,
may cause endless trouble. Try an iron shim and if this does not
work a strong clamp bolted on the stem above the tappet may
help.

Millmen often open the mortar when setting a tappet to put it
at the right height. Any millman that cannot set his tappets

from the cam-shaft floor had better be given his time, for it is easy enough to measure the distance on the stem for any height of drop needed and set the tappet accordingly.

Time is often lost by necessary repairs to the cushion on the feeder arm. Usually the end of the arm where the feed stem collar hits is horseshoe shaped and a piece of leather rivetted on. With this shaped arm it is best to make a complete ring of leather and allow it to rest loose on the arm. This will last for months without any attention. When the arm is straight it is best to bolt or rivet on an inch strip of wood with a cushion of belting nailed on the wood.

CHAPTER IV

LIMITATIONS OF THE SINGLE UNIT AND FIVE STAMP UNIT BATTERIES

This subject is of such supreme importance and enters so largely into the economical use of the stamp mill that I have thought it well to devote a separate chapter to its consideration. We have very little data of any comparisons of the action of these two classes of batteries on the same ore and what data we have is very conflicting. The trouble is that much of our technical literature is written by men who have made a mark in the profession and are then employed by some machinery house. This sort of information is useless no matter how well known the author may be.

The milling of gold ores is an art that has had the attention of many bright minds in the last fifty years. The choice of machinery for milling a gold or silver ore depends upon the chemical and physical nature of the precious metals in the ore as well as the nature of the enclosing rock. The problem usually rests upon the ability of the precious metals to form an amalgam with mercury and failing this, its affinity for some chemical solvent such as cyanide of potassium. The choice of machinery may be influenced by the amount of moisture in the ore, the hardness of the rock or the difficulty of separating the gold from the waste rock.

As our idea is to form some idea of the limitations of the two classes of stamp mills now in general use; the one and five stamp units, we will leave out of consideration other mechanical contrivances for breaking ore and consider that the rock has been delivered to the stamps at about 1 1/2 in. cube.

Amalgamators may be divided into two broad schools, those who claim that inside amalgamation at all times is a farce, and those who claim that in most cases when not followed by a chemical process it is necessary to amalgamate inside the mortar, on plates or otherwise if the best extraction is to be made. For the former crushing with rolls or single unit stamps is sufficient for all they require is the crushed ore to amalgamate on the outside plates. The latter desire units of five stamps with inside amalgamation. The latter claim that the mercury should should have a longer time to mix with the particles of gold than that possible when the only contact the gold has with the mercury is on the outside plates. On this difference of opinion and its practical results has hung the fate of many mining enterprises.

It is a noticeable fact that where amalgamation is done solely on the outside plates there will be found a cyanide annex, and it is necessary, unless in special cases such as for example in a mill where the principal values are in the concentrates and amalgamation is secondary or not necessary.

It is impossible to amalgamate inside the single unit batteries as at present made and they must be considered only as crushing machines. The screen surface being so great and the inside area other than that occupied by the stamp being so limited, any particle that will pass the screen is at once thrown out. This means that any mercury fed into the mortar is immediately thrown out before it has had time to mix with the ore. This is different in a five-stamp battery, even if the width at the discharge level is only 12 in. The mercury has time to mix with the ore and gathers more or less gold before being thrown through the screen.

If one submits a sample of ore for an amalgamating test it is ground with mercury or the ground ore mixed with mercury, the pulp is diluted with water and the mercury collected. This will represent the amount of gold recoverable by amalgamation. In one sense it does, but it does not represent the amount of gold recoverable by amalgamation with a single-unit mill. Why not duplicate the action that takes place in practice? Make two tests, the first as above outlined then running the pulp over, 10 ft. of amalgamated plates to recover the amalgam and the other by running the crushed ore over the same plate without having previously mixed it with mercury. If the results are the same there is no question of the preference being in favor of the single unit battery; if different, it is an arithmetical problem, and easily solved.

There is no doubt that many ores plate easily outside the mortar and there are many that do not. To expect fine gold or gold that may be coated with some objectionable material to amalgamate by simply washing over a sensitive plate is to expect too much. This gold must be well mixed with mercury and held in the battery long enough to be scoured and this result can only be obtained inside a five-stamp unit battery. The usefulness of this system has its limitations because the single unit batteries can crush ore at a less expense and the cyanide process may be able to take care of what gold escapes amalgamation.

The preference is a question of profits which means definite experiments with a given ore. Because a machine can crush ore at a less expense than other it is little recommendation, unless the ultimate result is a gain.

The tests necessary to design a mill for a particular ore are:

1. Amalgamation by thorough mixture, washing the pulp over 10 or 16 ft. of amalgamated plate.

2. Amalgamation by washing the crushed ore over the same amount of an amalgamated surface without having previously mixed the ore with mercury.

3. Concentration of the tailings from 1 and 2.

4. Cyanide extraction from tailings of 1, 2 and 3.

To reason inductively and arrive at some generalizations we must assume some figures to start with, the figures being what experience has proven to be reasonable.

First we assume that the single unit mill will crush 4 tons per stamp per day, while the five-stamp unit will crush 3 tons per stamp in the same time with the same power. Also that the saving of gold with the five-stamp unit battery will be 85 per cent. of the assay value. The question is, then, what must be the percentage of saving, using the single unit battery and cyanide plant, the cyanide recovery in both cases being the same providing the tailings from the five-stamp unit contains sufficient gold to make a cyanide plant pay. If the two types of mills make the same recovery the preference would be for that crushing at the least expense, but such is not the case. It is doubtful whether the tailings from the single unit mill will give as high a cyanide extraction as that from the five stamps, but we will assume that it does. My reason for stating this is that taken for granted the cyanide plant is arranged for the treatment of both sand and slimes the coarser the product going to the tanks the less will be the extraction and as all advocates of the single unit claim that it causes less slime than the five-stamp units the coarser product will give a less cyanide extraction as a rule.

If we tabulate the extraction and profits for ores of different values we will find:

1. To obtain the same profit per ton of ore crushed without a cyanide annex, assuming the crushing capacity as 4 to 3 the single unit need only amalgamate 60 per cent. of a $5.00 ore when the five-stamp unit must amalgamate 85 per cent. As the value of the ore increases so does the necessary extraction with the single unit batteries. If we assume a cyanide plant takes care of the tailings and the extraction in both cases is 90 per cent. of the residium we have the same comparative results. This shows that for small mills where the ore must be of a high grade to pay at all it is important to amalgamate well and save all that is possible without a cyanide plant. It is the small mill owner who usually buys the single-unit mill for he is attracted by the statement of the crushing capacity and does not consider the amalgamating ability of the machine.

2. The lower the cyanide extraction the higher must be the amalgamating efficiency of the single-unit mills. Should the cyanide extraction be different, for example, where it is 90 per cent. for the large unit and 70 per cent. for the single unit the usefulness of the latter has its limitations.

3. When we consider purely a concentrating ore the advantages are all in favor of the single-unit batteries, costing less per ton of ore crushed and a better product for concentrating.

If we consider some concrete examples the idea will become more apparent. Suppose we have made our preliminary tests and we have them tabulated. We are now ready to consider whether we will have a five-stamp or a single-unit mill.

No. 1. A quartz ore that by test No. 1 amalgamates 42 per cent., by test No. 2, 25 per cent. Test No. 3 shows 91 per cent. in both cases. The cyanide extraction is so low that it is out of the question. The cost of milling with the single unit is estimated at $1.10 and with the five-stamp unit $1.45 per ton. The ore averages $6.00 per ton.

This shows a profit of 25 cents a ton in favor of the single unit.

No. 2. An ore giving by test No. 1, 83 per cent., and test No. 2, 68 per cent. The extraction by concentration is 56 and 61 per cent. and the cyanide extraction in both cases is 84 per cent. The cost of milling is $1.40 and $1.05 per ton and the cyanide treatment $1.35 in both cases.   The ore averages $12.00 per ton.

```
Single  unit, Amalgamating .....................  $8.16
             Concentrating......................   2.34
             Cyanide.............................   1.26
                     Total ....................   11.76
                     Cost .....................    2.40
                     Profit.....................  $9.36

Five stamp unit, Amalgamating........................  $9.96
             Concentrating ........................    1.14
             Cyanide...............................    ....
                     Total.......................  $11.10
                     Cost........................    1.40
                     Profit......................   $9.70
```

In this case there is not enough left in the tailings of the five stamp unit to pay for cyaniding and as the process gives a better profit without a cyanide plant than the single unit mill gives with the cyanide addition, a five stamp unit mill is preferred. This is a sample of many typical cases.

Ex. No. 3. An ore amalgamating 5 per cent. in both cases, concentrating 1 per cent. of the residual values, but the whole tailing when reground to 150 mesh at an extra expense of 30 cents a ton gives an extraction of 81 per cent.   The ore averages $18.00 a ton.   The cost of milling is $1.45 and $1.95, plus in each case 30 cents a ton for regrinding.   Cyanide treatment, $1.55.

It will be seen at a glance that it will not pay to concentrate

```
Extraction by single unit, amalgamating.................  $1.50
             Concentrating..............................   4.10
                     Total..............................   5.60
                     Cost...............................   1.10
                     Profit.............................   4.50
Extraction by five-stamp unit, amalgamating............  $2.50
             Concentrating..............................   3.20
                     Total..............................   5.70
                     Cost..............................    1.45
                     Profit.............................  $4.25
```

and that the balance will be in favor of that process which crushes at the least expense, namely, the single-unit battery.

Makers of single-unit mills are apt to claim too much for their machinery and make statements about the five-stamp units that do not represent a fair comparison.   For example, to compare the amount of ore crushed in a five-stamp amalgamating mortar in a mill that has been running on worn shoes and dies with a single unit mortar with new shoes and dies is not fair. Why not compare a five-stamp double discharge narrow mortar with a single-unit battery, both using new shoes and dies?

As the Nissen stamp is one of the best makes of single-unit mills a description of the 100-stamp mill of the Boston Con. Co., is here given.   "The battery frames are of wood.   The stamps

Fig. 13.—Nissen mortar.

are mounted in batteries of eight, further divided by a center post into two groups of four stamps each.   Each Nissen stamp, of course, has a separate mortar.   The frame consists of two 14 × 26 in. end posts, one 20 × 26 in. center post, and two 14 × 26 in. jack posts, resting on a 14 × 20 in. over-hanging jack beam, and two 14 × 16 in. girts.   Each post is braced against the battery bin by wooden beams resting in brackers bolted to the bin posts.   Each set of four stamps has its own cam-shaft, 6 in. in diameter, and driving pulley, 72 in. in diameter.   Consequently each center post carries a double bearing.   These bearings are used without caps.   Each set of twenty stamps is driven by a 75 h. p. motor.   These motors are placed below the bin and behind the battery being mounted on the bin framework where they are easily accessible. The driving belts to each group of four stamps is 14 in. wide."

"The cams are fastened to the cam-shafts by Candan fasteners, which are somewhat similar to the Blanton fasteners. The tappets and cams are made of chrome steel; the stems run in cast iron guides. The cam-shafts are made of malleable iron. The shoes and dies are made of white cast iron, cast at the company's foundry at the plant."

"The mortars have a specially heavy base so that no anvil block is required. A rubber sheet 1/2 an inch thick, is placed under the mortars and then the mortars bolted down to the concrete mortar-blocks. The stamps weigh 1500 lb. and make

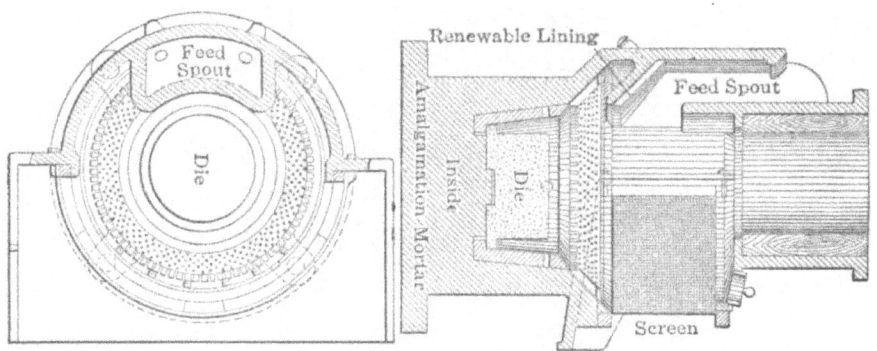

Fig. 14.—Nissen amalgamating mortar.

110 drops, 7 in. high, each minute. A screen analysis on the pulp from the battery shows 12.5 per cent. on a fifty-mesh screen, 14.5 per cent. on eighty mesh, 9.5 per cent. on 100-mesh, 17 per cent. on 150-mesh, 1 per cent. on 200-mesh and 45.5 per cent. through a 200-mesh."

"The batteries are fitted with diagonal punched screens equilalent to twenty-six mesh, No. 26 wire screen. The average size of the ore going to the battery is 5/8 in. About 10 gal. of water a minute are fed to each stamp, and the duty is about 8.3 tons per stamp. It is hoped to increase this duty to 10 tons by increasing the weight of the stamps 200 lb."

Illustrations 13 and 14 show the Nissen stamp arranged for amalgamation inside the mortar.

CHAPTER V

HISTORY OF AMALGAMATION

The fact that mercury, when brought into contact with gold or silver immediately forms a combination with them, called amalgam, was known and utilized in the most ancient times.

This circumstance is clearly attested by the evidence presented. It is necessary to call attention to the fact that there is a rather complicated metallurgical operation known as the "patio process," patio being the Spanish name for the court or the enclosure in which work is carried on. In the patio process the pulverized ores of silver, after being spread out over the court in a layer of even thickness, receive a charge consisting of mercury, common salt and bluestone, water being added in the requisite amount. This combination is called "magistral." The stirring of this mixture results in converting the argentiferous portion of the ore into a condition which causes it to form an amalgam with the mercury.

While there is nothing to show that the ancients were familiar with this or some similar process, there is reason to believe that they understood the simple amalgamation of gold and doubtless of some silver ores. At the present day the natives in many districts of India employ this process and it is believed that the practice has continued since the Moslem conquest of India in the tenth century.

So early as 300 B. C. Theophrastus mentions "liquid silver" which he says was obtained by rubbing cinnabar with vinegar in a copper vessel. Dioscorides describes the production of mercury by subliming with charcoal in an iron pot.[1]

The mines of Almaden according to Pliny supplied the Greeks with red cinnabar 700 years before the Christian era; and Rome in his time received great quantities from the same mines.

Mercury and some of its uses appear to have been known at a very early period of the world's history. Aristotle relates that Daedalus gave motion to a wooden Venus by means of a current of mercury.

[1] "New International Encyclopedia."

Dioscorides describes a method of its reduction from cinnabar.[1]

Sir Gardiner Wilkinson says that mercury in flasks has been found in ancient Egyptian tombs. The metal was sacred to Osiris. The discovery of mercury in the ancient tombs makes it probable that the Eygptians used it for recovering the precious metals in connection with the stone tables over which they washed the crushed gold ore.

In the Greek mining district of Laurium were deposits containing cinnabar (the sulphide of mercury) and copper, the former of which was worked at a profit. Near at hand were their silver veins. The close proximity of the three essential metals of which later became the patio process of the Spaniards could scarcely have failed to have directed the attention of the ancient Greeks to the beneficent operation of what we now call "Magistral." It is quite likely, however, that the Greeks employed simple amalgamation for the Romans, who borrowed the Greek civilization, were familiar with it. It is mentioned by Vitruvius about 27 B. C. and described by Pliny about A. D. 73 The latter says "Gold is the only substance that quicksilver attracts to itself; hence it is, that it is such an excellent refiner of gold, for on being shaken in an earthen vessel with gold it rejects all dross, clinging only to the gold; the dross being expelled, it only remains to separate the quicksilver from the gold. This is done by enclosing the mixture (amalgam) in a well prepared skin, which being squeezed, exudes the quicksilver, like a sort of perspiration, leaving the pure gold behind."

This is the simple process resorted to by miners and millmen to this day, the "skin" used being a piece of chamois or as a substitute, a bag of drilling or canvas.

Edrisi—Twelfth Century—informs us that "during the eleventh century and long previously," quicksilver amalgamation was practised by the Arabians and the negro miners of Western Africa, Abyssinia, etc.

Amalgamation was employed by the gold miners of Portugal during the reign of King Diniz, A. D. 1279–1323. It had probably been learned from the Arabs. In A. D. 1525. a miner named Paolo Belvio, was sent with a "provision of quicksilver" to Hayti, in order to expedite the gold washing there by means of amalgamation.

Herrera, in his history of Spanish America, says that, although

[1] "Encyclopedia of Chemistry."

previous to the opening of the mines of Potosi, the Spaniards were familiar with the art of amalgamating gold, they were unaware that mercury would amalgamate with silver. What he may have meant was that previous to the opening of the Mexican silver mines by the Spaniards, they were familiar with the simple amalgamation of both gold and silver, but not with the complicated patio process, which may have been the discovery of Medina, as has been asserted in 1557, and that in1571 Pedro Fernandez de Valasco carried this process to Potosi, in Bolivia.[1]

Agricola, who was well acquainted with simple amalgamation, and also with the reduction of silver ores by roasting, may have suggested the basis of the process.

There is little doubt the success of the so-called "transmutation" of mercury or silver into gold, as practised by the alchemists of the Middle Ages, was wholly due to the gold existing in the amalgam of silver which was extensively used. It has always been, and is now, the common impression that gold after amalgamation may be separated from mercury by simply squeezing through chamois skin or a leather bag, or a thick piece of canvas. Pure gold, finely divided in the form of powder when amalgamated with mercury in the proportion of over one of gold to 200 of mercury, will go right through a chamois skin, but, if a small quantity of silver be present, will remain behind. Now it would be easy for a charlatan, of the older times to squeeze this mixture of pure gold amalgam through many skins and then pretend to "transmute" some other metal into gold by using this contaminated mercury.

Following is a receipt of "How true and perfect gold may be made by art without loss to the workman," as written in the year 1738.

"I took 8 oz. of regulus of iron and copper and 16 oz. of common sublimate and made these ingredients into a fine powder, and so put them into a glass retort, and drew from them first an oyl, then a substance like butter, and lastly, a yellow sublimate, tincted with the tincture of iron and copper, which sublimate I rectified three or four times till it was pure, then I mixed it with equal parts of an amalgam of silver and quicksilver and put it into another glass retort. I retorted it for a week. It then melted the silver, refined it and found gold."

[1] Del Mar's "History of the Precious Metals."

So late as 1865 Overman wrote: "Amalgamation is a process not adopted to our social conditions. It is too laborious to secure success in ordinary cases of common or average ore." He perhaps had in mind the patio process or the pan-amalgamation process then in use.

When we remember that amalgamation was so little understood in his day, that in cases, urea was used in the pans, it is no wonder he made the above remark.

In 1868 Urbas recommended the employment of hydraulic pressure in the amalgamation process, and Wycoff advised that finely-divided ore should be boiled under pressure with half its weight of mercury.

The amalgamation of gold and silver ores in the sixties and seventies was considered as a secret art and has been so to a certain extent, until lately. The technical journals have been the means through which this art has become public property and all the improvements in this art have been due to millmen expressing their ideas through this medium.

CHAPTER VI

AMALGAMATION

The first and foremost duty of every amalgamator should be to see that no free gold, amalgam or mercury goes over the end of the amalgamated plates. As a rule the free gold and amalgam which finds its way to the cyanide works is difficult to recover and the amalgamator should see that as little as possible goes that way. It would be well if the amalgamator could put the cyanide chemist out of business and he could do it in many cases by more attention to details.

All millmen employ the same general principles and work toward a common goal, but which they reach by divers routes and with varying degrees of success. Each man has his own particular process—perhaps a secret one—which he employs with supreme confidence in its efficacy, while exhibiting an equal degree of suspicions of the methods of others. Perhaps if the underlying principles of amalgamation were better understood, the present lack of uniformity in the methods now employed would disappear. In order to grasp the subject thoroughly, it is necessary to understand the chemical and physical properties of mercury and amalgams.

The specific gravity of mercury varies from 13.55 to 13.59, according to the temperature. It congeals at —39° F., or —40° C., and may be then beaten out like any other metal. At 662° F. it boils and distils off a vapor which being condensed forms mercury as we usually see it.

To obtain pure mercury for experimental purposes, sublime mercuric chloride with one part iron filings in a glass or earthen retort. Too often experiments are made with impure mercury and consequently the results are misleading.

The mercury of commerce is usually contaminated with variable quantities of lead, zinc, tin, bismuth and similar metals; while mercury that has been retorted in a mill may contain in addition to the above arsenic, antimony and volatile carbonaceous compounds. An impure mercury may be purified by any of several processes, of which the following has been recommended for its simplicity and the excellent result it gives. Place the mercury in an ordinary retort with about one-tenth its weight of cinnabar and distill very slowly. The impure metals will combine with the sulphur in the cinnabar to form sulphides and will be found in the bottom of the retort when opened. Retorting at a high temperature is always productive of inferior quicksilver. It is done to save time and insure the expulsion of all the mercury from the retort, but it is worth while to make haste slowly in work of this kind.

To ascertain the purity of the metal two methods may be used, one chemical, the other physical. In the former a quantity of the metal is dissolved in nitric acid, evaporated, heated to redness or fused with pure sulphur. In either case nothing should remain if the metal is pure, but if there is residue the impurities are proportionate to the weights. In the latter, when mercury is dropped upon a smooth, but slightly inclined surface, it breaks up into drops which retain the spherical form, but if contaminated with tin, lead or such adulterants or impurities the form of the globule will appear elongated and will leave a "tail" of impurities.

Bianchi recommends boiling impure mercury with one-eighty-seventh of its weight of nitrate of mercury dissolved in water or with a small quantity of nitric acid. The nitric acid or the acid radical in the nitrate converts the base metals into nitrates which are all soluble in water and therefore easily washed off.

Ulex purified mercury by agitating with one-thirty-second of its weight of ferric chloride; after ten minutes shaking, the mercury is washed with water, dried and the finely divided globules reunited by the application of a gentle heat.

Mercury may be purified under certain conditions by electrolysis where a chloride or an aqueous oxyhydrate is dissociated, forming free chlorine or hydrogen. The conditions are that the electrolytic action does not throw down base metals as chlorides or hydrates or that the chlorine does not form injurious substances with the constituents of the ore, such for example, as sulphuretted hydrogen, which we know will form sulphides.

The metal evaporates at common temperatures as well as *in vacuo*. This was proved by Farraday by suspending gold leaf in a flask containing some of the metal, when, after a few weeks, the lower portion of the leaf appeared amalgamated. Mergent by using a test paper coated with ammoniacal silver nitrate proved evaporation of the metal at as low as — 44° F., the volatilization being uninterrupted even by congelation.

Pure mercury is not acted upon when exposed to air, oxygen, nitrous or nitric oxide, or carbonic acid gases. Kept in a temperature approaching boiling in contact with air it oxidizes slowly. In damp air it gradually becomes coated with an oxide. Impure mercury becomes coated with a film of oxide even in dry air.

If shaken with water, ether, oil of turpentine or fatty unctuous matters it loses its metallic appearance, in which the metal is not altered but deadened—that is, reduced to minute and isolated globules, in which state it is easily washed away in suspension.

Sulphuric acid when diluted does not attack mercury even when aided by heat, but concentrated acid at a high temperature readily converts it into mercurous sulphate, sulphurous acid being given off.

Hydrochloric acid does not attack mercury, but chlorides in a gaseous condition or in a solution of water do.

Nitric acid even when diluted dissolves mercury, forming a nitrate.

Mercury poured upon dry bodies does not wet them like other liquids, but flows off in drops, except in the case of metals with which it forms amalgams, uniting with facility with many metals as well as sulphur, phosphorus, chlorine, bromine and iodine.

Gold amalgam containing 90 per cent. mercury is fluid and with 87 1/2 per cent. it is pasty.

Solutions of pure gold amalgam in mercury when pressed through chamois leather leave a residue containing about 66 per cent. gold, the mercury passing through the leather bears a little gold, the amount depending upon the temperature. Then when at 0°, it is 0.11 per cent., at 20° 0.126 and at 100° 0.65 per cent.

The gold bearing mercury can be decomposed by heat into gold and mercury. Mercury thus distilled carries a little gold with it, under ordinary conditions about 0.1 per cent.

Amalgams take a crystalline form and with different degrees of dilution, different forms, thus indicating combinations in definite proportions or a chemical mixture. This is disputed by those who hold that an amalgam is a mechanical mixture.

When mercury contains dissolved in it lead, zinc or other

extraneous metals it tarnishes rapidly, loses its perfect liquidity and is unfit for practical application unless purified. This can be done in mill practice by covering the surface of the metal placed in a shallow dish, with dilute nitric acid, stirring frequently and washing out the impurities.

If a film of oxide of a metal adheres to the mercury it may be removed by agitating it violently in a bottle with some powdered sugar, then blowing air into the bottle, repeating the shaking and blowing several times, then filtering.[1]

The presence of sulphur (pyrite) is a frequent cause of interference with amalgamation, causing the formation of a sulphide.

According to Bettel mercury may be purified by using a 2 per cent. solution of cyanide of potash with the addition of a little Na_2O_2. After an interval the gold in the amalgam dissolves as well as the impurities, so care must be taken to stop the action at the right time.

Practical Application of Amalgamation

Clean mercury, clean particles of metal to amalgamate and intimate contact, these are the three fundamental requisites of successful amalgamation. The character of the crushing and subsequent handling of the crushed particles determine the two latter requisites, but clean mercury or clean amalgam is the prime factor, and to obtain it, requires only a little care and thought. Careful metallurgists will neither use impure mercury for experimental tests nor foul mercury in practical work. To many, retorted mercury is clean. This is not necessarily so, as has been shown.

Clean or unclean particles of metal, as understood in the preceding paragraph may mean that the enclosing rock must be broken sufficiently fine to free the metallic particles so that the mercury when brought into contact will completely surround it, or it may mean that the metallic particle being free is covered with some foreign substance that prevents the mercury from wetting the particle and so holding it. The character of the crushing machinery determines to a great extent the state of this particle; if the gold is clean, simply crushing to the desired mesh is sufficient, if unclean, the particle must be crushed in a manner that will rub off the adhering substance or some

[1] "American Encyclopedia."

chemical reaction must take place that will dissolve this substance. Here is possibly the only field for electrolytic amalgamating appliances.

The subsequent handling of the crushed particle determines the manner and length of time the metal comes in contact with the mercury. The only machine that can regulate this to a nicety is the stamp mill, although many other machines are good amalgamators. The choice of the inside shape of the mortar, the number and position of the inside plates and the height at which the screen opening is placed, the weight and height of the drop of the stamp, the number of drops per minute, the order of drop, the amount and character of the feed water, the size of the screen openings and the amount and character of feeding the mercury all influence the number of times the particle of metal comes into contact with the mercury. It is a mathematical certainty that the more often a particle of metal comes in contact with any surface that has the ability to hold it the more likely it is to be held by that surface, especially if the particle and surface are clean. For example, it is not contended that a particle crushed to the size of sixty mesh, if dashed against an amalgamated plate fifty times, is more liable to be caught than the same particle crushed only to thirty mesh, for the finely crushed particle may not be physically in such good condition as the coarser particle, but it is contended that when the particle has been crushed to its most efficient size the more often contact takes place, the more likelihood of its being caught.

Whether it is economical to hold that particle long enough for a successful catch and so limit the number of such particles freed is another question, and one that must be determined by experiment. When an amalgamator says he can catch as much gold on the outside plates alone as with inside amalgamating with or without plates he is not stating a fact, and probably means that by amalgamating on the outside plates entirely he can put so much more ore through the mortar that any loss in gold is more than made up by the amount of ore crushed and consequently the bullion produced in a given time.

Here is where we see the application of this fact to the style of mill desired for a particular ore. If I can catch as much gold on my outside plates without inside amalgamation as I can with inside plates, I am going to crush as much ore as possible, therefore I prefer a fast crushing single unit or doubled discharge narrow mortar five-stamp unit if per contra, I prefer roomy inside amalgamating mortars providing of course there is no after process. If the cyanide process follows the mill, the cyanide process used will determine to some extent the preference. The following dissertation of inside amalgamation is from K. C. Parrish.

"Inside amalgamation is often advisable when there is difficulty with the outside plates on which a scum sometimes forms in spite of all precautions. Amalgam is much safer and less apt to be lost in ordinary or careless work when caught on the outside, but there are some cases where gold can be caught on the inside when it cannot be amalgamated on the outside."

"Several years ago I had occasion to make some mill runs on several hundred tons of high-grade ore in Columbia, South America. The ore contained about 13 per cent. of sulphides, half of which was pyrite and the rest zinc and lead sulphides in about equal proportions. It was found that a dark colored scum formed on the outside plates within a few minutes after starting the mill; the addition of chemicals did not help matters. The outside plates were placed very steep to clear themselves of the sulphides as quickly as possible. Frequent and careful dressing of the silver apron plates had no special effect; different degrees of hardness, cleaner mercury, wells and traps did not improve the extraction."

"The gold was mostly fine and part of it had to be brightened before it would amalgamate. This was done by using a higher discharge and finer screen. It was noticed that if the gold did not amalgamate on the inside but a small proportion of it could be caught on the outside plates. The outside plates could not be kept free from the scum and even the amalgam that was put on or occasionally caught, would sicken and scour off. Careful watching and feeling of the plates and the more frequent addition of quicksilver (as often as every fifteen minutes) raised the extraction from a total of about 50 to 70 per cent."

"The advantage of inside-plate amalgamation is that the plates are always kept clean and bright by the splash of the pulp. Sufficient quicksilver must be added to the mortar to take up all the gold, but not so much as to soften the plates. It is best added frequently in small amounts."

In my opinion the shape and area of the inside of the mortar is as important in considering our California ores as the position of the outside plates. We do not want to use cyanide if possible. The process is expensive to work and the results often disappointing.

Copper, because of its affinity for mercury and the fact that it absorbs mercury without changing its form, has been universally used as a surface on which to amalgamate the precious metals. Muntz metal, an alloy of 60 per cent. copper and 40 per cent. zinc, has been tried and although satisfactory in some cases has not been generally adopted. This alloy forms an electric couple which may help or retard the efficiency of the mercury according to the constituents of the ore.

The California practice is to use electro-plated copper plates with from 1 to 3 oz. of silver per square foot of surface. We have our reasons for so doing. Our ores are low grade, we make a better percentage of extraction than with the plain copper as usually treated. We cannot afford to lock up the quantity of gold necessary to keep plain copper in good condition. Few of our mills have a cyanide annex, consequently the plates must be at their maximum efficiency all the time, a state possible with copper, but not probable. No doubt a good soft copper plate well amalgamated, that is completely covered with a layer of gold or silver amalgam, is as good as a silver-plated copper with the same amalgam, but after cleaning it takes time to get it back to its previous condition, whereas the silver-plated plate is ready at once.

It is our custom to use soft copper for inside, splash and lip plates, but these are as often silver-plated. The inside plates being nearest the source of crushing have the first and best opportunity of catching the freed gold. The amount of gold caught on the inside plates together with the amalgam in the mortar between and about the dies usually amounts to about half the total amalgamated gold.

The South African practice is to use the plain soft copper plates. The reason is plain. The results are as good once the plate is "set" with a good coating of amalgam, and they can afford to allow this coating to remain on the plate. The cyanide plant in all cases follows the plates, and any loss due to poor amalgamation is taken care of. After each clean-up day the plates are in poor condition for a few days at least, unless a great quantity of amalgam is rubbed in and the amalgamator spends much of his time on the plates.

A bare spot on a silvered plate is more difficult to cover and to keep covered than one on plain copper. The electro-plating of a sheet of copper changes its molecular structure so that though originally soft and easily amalgamated it has become more closely grained, and the plate will not absorb mercury readily. When loosely amalgamated it soon wears bare again. The only remedy is to amalgamate well and often with amalgam, until the amalgam has set and holds firmly, or failing this to have the plate resilvered. All that chemicals can do is to prepare the copper so that the amalgam will form a close union. Sometimes by heating the plate and plunging in cold water the copper may be improved by softening.

To prepare plain copper plates for amalgamation the copper should be thoroughly scoured with fine sand until bright, then mercury rubbed in with an application of cyanide of potassium and sal ammoniac, or a little nitrate of mercury will hasten the action. The plate now being amalgamated will tarnish if exposed to the air. To prevent this and form a good surface for amalgamating gold, silver amalgam should be rubbed in, given sufficient time to soak in and at intervals other lots of amalgam rubbed in. The amount of silver necessary to produce a good surface may be half an ounce to the square foot. Should a bare spot appear on this plate it should be scoured until bright and silver or gold amalgam rubbed in. One has often noticed that when amalgam is squeezed through chamois skin and the mercury allowed to stand for some time a quantity of thick fine grained amalgam settles to the bottom. This when rubbed on plates makes fine, smooth coating, and being the fine gold that filters through the leather is absorbed by the copper more readily than coarse grained amalgam. This applies to the silver amalgam which should be prepared so that it is soft and smooth.

Electro-plated coppers must be treated carefully at all times. They should never be scoured, treated with a chisel or any chemicals used if it is possible to do without. Cyanide of potash, lye, caustic soda and all acids have an injurious effect. The silver may be dissolved or the plate rendered hard and glassy and not amalgamating easily as well as readily fouled. The plates should be protected against ill-usage, such as affording a rest for heavy machinery or a slide for the millman. Sand should not be piled up or rubbed on the plate, for a bare spot once started will spread unless immediately covered and set with amalgam. To prevent the carelessness usual in most mills the tables should not be set next to the mortars and if they are the tables should be on wheels so that they can be pushed away from the mortars when handling shoes and dies. Plates away from the mortars would not be subject to the outrage due to the vibrations of the battery frame.

The following is recommended by a South African engineer and is an excellent idea: "When changing screens, working at shoes and dies, etc., there is always far too much scratching and tearing up of the plates and amalgam by throwing tools, screen pins, pulling coarse gravel out of the mortar box, etc., on the plates. This should all be avoided and can be with a little care. Before opening a mortar box a rubber apron should be first laid down over the plate. There should be two aprons kept, one small and one large; the small apron to be used when changing screens, and the large one used when working on shoes, dies, etc. The apron should be laid down before the shoe and die board is put on the table. A roller should be attached to each apron, so that it can be rolled up after use." The rubber aprons should be made to fit close up to the lip of the mortar box, and close to the sides of the table. There should also be a hollow tray, so made that it will fit close to and against the lip of the mortar box, and run along the full length of box. The tray should have several drainage holes in it.

"To change a screen, when the stamps are hung up, and the water shut off, lay down the smaller of the two rubber aprons, then put the tray in position. The screen pins, splash board, screen and tools can all be laid on the apron without damaging the plate or amalgam. When hosing the gravel off the lip of the mortar box, the tray should be held tightly in position, so that it catch the gravel and coarse sand. The new screen having been adjusted, the apron should be swept, and the sweepings put in the tray. The contents of the tray may be returned to the mortar at once."

If grease fouls the plate the use of a mineral soap, such as borax soap is preferable to cyanide of potash or lye. A bare spot may be covered as outlined above by rubbing in silver amalgam, or this may be prepared directly on the plate by an application of a weak solution of cyanide of potash in which some nitrate of silver has been dissolved.

We have all made certain observations which guide us in our

work, and while some of these are self-evident from the character of the materials we employ, others come to us by observation. The first and foremost is no doubt familiar to many—the fact that amalgam of the right consistence holds gold better than mercury alone. From this fact arises all the failures of masses of mercury whether in pools or sheets, to effect a saving equal to plate amalgamation.

One inventor has a continuously falling sheet of mercury against which the pulp is circulated. He claims that this will catch gold that cannot be found by assay. The fact is the amalgamated copper plate will do better work, because the amalgam on its surface is a better catcher of gold than the mercury alone. I am not condemning mercury traps, for these are a necessary adjunct of the plate amalgamator.

Another observation is that an amalgam may be diluted on a plate subject to the ordinary vibrations so that a separation of mercury and the precious metals takes place—the mercury sliding down the plate, the metallic amalgam remaining at the head of the plate, hard and not easily removed. Amalgam may be kept soft enough for any reasonable system of amalgamation without running, but should a certain point in the dilution be passed, separation takes place. This is not an idle observation, but one of repeated experience.

Another observation is that when soft amalgam is thrown against an amalgamated surface a separation takes place just as if squeezed through leather. Here is a reason for our inside plates, the churning inside the mortar dashes the gold or amalgam against the plate where it sticks just as a spit-ball will on a wall, and yet there is mercury going through the screen. We may, however, dash too much mercury against these plates and produce a soft plate from which the accumulated amalgam will scour. Experience teaches us the limit. This peculiarity also indicates the use of drops from plates to plate, not only to turn the pulp over, but to allow it to fall with a certain amount of force. This is only allowed in practice with plates that have a thick coating of amalgam, such as lip or splash plates. Any drop on the table plates should be under 1 in., for the layer of amalgam being thin is easily scoured.

A perfect contact of gold particles with the amalgamated surface is obtained when every particle of gold is brought into contact as many times as possible under ideal conditions. These ideal conditions may vary with the physical character of the gold particles or the enclosing rock, but certain conditions apply generally. These are that the rock be broken fine enough to free the gold and no finer, that it flow over the amalgamated surface in a thin layer, but not so thin that air may come in contact with the gold and cause flotation, that it move slowly over the surface, that it be thoroughly mixed and amalgamated before it comes in contact with the plates and that no substance be allowed to touch either the gold or the mercury that may hinder union.

Some ores contain ingredients that require correction or neutralization. An acid ore may require the addition of lime, an antimony ore, some chemical such as salt to form a new compound not so injurious or an electric current to produce a different combination.

Besides being affected by foreign substances the quality of amalgamation may be affected by the conditions of crushing inside the mortar, size of mesh, the number and position of the plates, the method of dressing the plates, the frequence of dressing the plates and the temperature and constituent chemical in the water.

The reason why a different method, in details at least, is used for different ores lies in the fact that the physical conditions of every rock and every particle of gold is different. In some rock the gold is flat, in some grained, in some amorphous, in some crystalline, sometimes the enclosing rock is hard and brittle, in other cases soft or flexible; in some the gold readily escapes when the rock is broken, in some it must be crushed fine, in some the gold is in small grains while in others it is coarse.

In preparing a scheme for amalgamation all these various factors must be considered. Is it any wonder then that there is no hard or fast rule to follow? The safest rule is to employ a millman who understands the character of the substances he is to employ and pay him sufficiently so that he takes interest in his work and uses his brains for other purposes, but to do and achieve as little as possible without being fired. A superintendent should be broad minded enough first to employ a competent man and keep him, allow him some latitude in his work, but keep him checked up all the time by the assayer who should have no interest in falsifying results. This last phrase may be uncalled for, but experience teaches otherwise. For example, an assayer who also has charge of the cyanide plant may find that by falsifying the millman's work, his own work may appear to better advantage.

We now return to the consideration of inside amalgamation by which I mean the feeding of mercury inside the mortar, for some of my friends contend that with any ore as much gold can be caught on the outside plates alone without inside amalgamation as with inside plates and feeding "quick" in the mortar. My friends, I believe you are wrong. My experience is otherwise. The quotation on a former page from K. C. Parrish is a repetition of my own experience. I need only cite one instance. We had two pairs of crushing rolls and a receptacle for mixing the crushed ore (30-mesh) with water, before running over the plates. This arrangement only amalgamated about 40 per cent. of the gold. The mill was later changed to a stamp mill with a five-stamp amalgamating battery and inside amalgamation. The saving was now over 80 per cent. I am not contending that this is the case with every ore, for the Yellow Aster discarded inside plates, although feeding quick in the mortars, with no loss as compared with the previous system, where inside plates were used, and I have also amalgamated ore crushed by rolls with tailings under 20 cents a ton.

The case of a California ore may be interesting. As every millman knows it is next to impossible to amalgamate with plates inside the usual single unit mortars, for the plates scour abnormally. The mill practice was to feed mercury inside and plate outside. The tailings went a little over a dollar. To determine what effect inside plates might have, new silvered plates were put inside and the sample of tailings kept separate. For a short time after these plates were used the values fell to 25 cents a ton to a trace and soon went back to the old figures when the plates had been scoured of silver and amalgam. This was tried many times with the same results. It seemed to me a clear proof that we needed amalgamating mortars, for the tailings were not rich enough to cyanide and a difference of more than the cost of milling could be saved.

With the single unit mills amalgamating on plates must be done outside the mortar. The loss of mercury will be inappreciable if amalgamation takes place entirely outside, but the loss in gold may more than compensate for this gain. With this system of amalgamating the mercury is fed on the plates in a fine spray and the plates should be brushed often—at least every six hours—for no matter how carefully the mercury is fed it does not distribute as evenly as when mercury is fed inside.

The California millman depends as a rule upon amalgamation and concentration to effect the saving of the precious metals, therefore he designs his mortar to catch gold as well as crush ore- in other words, we do not aim to crush faster than we can amalgamate. This does not tend to promote cheap crushing, but it is commercially right. We seldom use stamps over 1050 lbs., but there is no reason why heavier stamps could not be used with a sufficient width at discharge to do good amalgamating.

The usual method is to feed mercury inside the mortar at short intervals, preferably every half hour, the amount fed depending on the condition of the outside plates. The plates nearest the

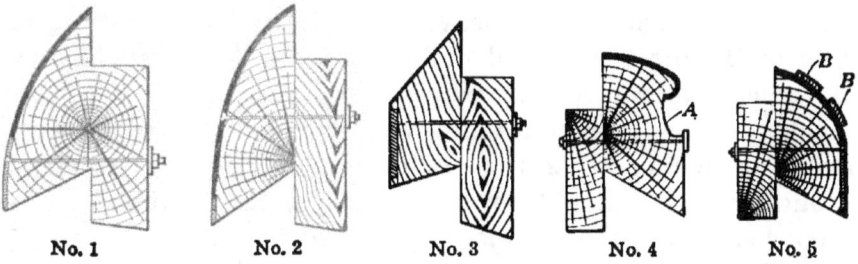

No. 1   No. 2   No. 3   No. 4   No. 5

FIG. 15.—Chuck blocks.

mortar—splash and lip plates—indicate the consistency of the amalgam on the inside plates if there any and for this reason they should not be touched except to clean off any foulness. If bare spots occur they should be amalgamated and the "feather" edge should be scraped off every day. The illustrations 1 to 5, Fig. 15, show different forms of inside chuck-block plates, the plates that are put directly under the screen frame. This with the plate placed behind the dies are referred to as inside plates. The back plates are usually 5 to 10 in. wide the full length of the inside of the mortar. The chuck-block plates are placed under the screen frame the full width of the mortar opening and may be of various shapes. No. 1 and 2 are practically identical and differ in that in No. 1 the block is solid and does not reach as far down toward the top of the die. No. 3. is the form usually recommended by one of the prominent iron works. No. 4 is recommended by C. O'Brien. The depression A catches and holds the gold without any fear of scouring. No. 5 is the form at use at the Yellow Aster Mill. Several iron bars are bolted on the plate. The gold accumulates between the bars BB which protect the plate and prevent scouring. The bars hold the copper firmly, allowing no chance of it breaking loose and getting under the stamps. By striking this plate with a hammer most of the amalgam will come off.

The table or battery plates are kept at the right consistency— usually putty-like, soft and bright. If the tendency is to harden and the inside plates tend to scour from being too soft then the battery should be hung up and the table plates softened with mercury. Inside plates may also scour from being too hard and brittle; the remedy is more mercury fed inside and less outside.

- If the grade of ore is kept somewhat uniform the millman soon learns to know how much mercury to use on his outside plates and at what consistency to leave them. Any sudden change of grade in the ore the careful millman soon notices and changes his "quick" allowance accordingly.

Silver ores and ores containing very fine gold require a greater amount of mercury than their value would indicate. Amalgam on outside plates should never be so dry as to require a chisel, except perhaps at the extreme top, nor so wet that the mercury flows off in "tears." Any intermediate condition may be successful. The union of mercury and gold is a mechanical one just as the union of a fly and molasses is mechanical and both act about in the same way. If the molasses be thick and pasty the fly remains, if thin the fly gradually works its way off. The fly uses its muscular strength, the gold particle the strength of the flowing pulp.

We may here discern one reason for keeping different parts of the plate of different consistencies when treating an ore that requires a thin amalgam which from change of temperature of the water or carelessness is apt to run, the lower portion of the plate may be kept stiff to absorb any of this overflow, or if contrariwise the plates are kept stiff as a rule, the lower portion of the plate may be kept wet to catch any particles escaping from above.

Plates should never vibrate, for it has the same effect as squeezing through leather. They may shake, which is a different motion entirely. We in California have not taken kindly to the shaking tables and why should we when we can do as well without? The illustration, Fig., 16 shows the shaking tables in use at the New Bantjes Mill, Transvaal.

Plates are dressed whenever necessary to keep the surface in good condition. Where outside amalgamation alone is practised it is better to dress often to correct any poor feeding of mercury. Most mill plates require dressing at least twice a day, once to clean off the excess of amalgam and once to clean and spread the amalgam evenly over the surface. When dressing the plates on the night shift, unless the ore is rich the object is

FIG. 16.—Shaking amalgamating tables.

simply to improve the amalgamating surface and take off any stray pieces of amalgam that may have become detached from the screen or upper plates. If the lower plate is in good condition one may begin at the upper plate and come down, first brushing up to the head and taking off any loose amalgam. Then brush vigorously until the plate is bright and the amalgam spread evenly, then with a whisk broom or soft brush drawn lightly from side to side. The object is not so much to make riffles as to lay any loose amalgam. If this is not done these pieces will surely be pushed down the plate into the trap which may or may not hold them. The usual procedure when dressing plates on the day shift, is to turn off the water and hang up the stamps. If the stamps are hung up first the battery fills with water, and when starting a flood of water pours over the plates taking with it any loose amalgam that might have remained if not so roughly treated.

The screen frame, splash and lip plates are washed off with the hose and the tables cleaned of all sand. At this point a trough may be placed at the end of the plate to catch any loose pieces of amalgam. The table plates are now sprinkled with mercury and rubbed with a brush. The splash plates are taken off, the screen frame freed of any amalgam and the edges of the splash and lip plate treated the same way. Any black or bare spots on these plates are brightened by rubbing in "quick" with perhaps the help of a chemical such as cyanide of potash, mercuric nitrate or chloride. The loose amalgam is all brushed up to the head and taken off. Any surplus of amalgam is taken off with the whisk broom alone or lightly with a rubber scraper. The plate is now well brushed, adding mercury in a fine spray until the amalgam is of the right consistency. Sometimes the inside plates have become too soft when it may be necessary to make the table plates wetter than usual and feed less mercury inside, or it may be that the inside plates have become too hard when it may be advisable to have the table plate harder than usual so that more mercury may be fed in the mortar. The whisk broom is now applied from side to side or transversely to lay any loose amalgam. The splash and lip plates are now returned and thoroughly hosed off, the box or trough at the end of the plate catching any loose amalgam. The battery is now started and the water turned on. Two good men can go through this process at the rate of ten minutes a plate and twenty plates have been done at the rate of seven minutes a plate, the two men looking after a 100 stamps at the same time.

Plates are dressed with a variety of chemicals according to the cause of the discoloration. The rule should be not to use chemicals unless absolutely necessary and when they must be used to make the solutions as weak as will do the work. For general purposes cyanide of potash or caustic soda are used say about a 1 per cent. solution or a solution may be made of both these chemicals. Where sulphurets stain the plates some substance must be used that will decompose the sulphurets such as chemicals that will produce a chloride and at the same time preserve a neutral solution. Sal ammoniac and carbonate of soda or sal ammoniac and lime may answer this purpose. Some recommend salt and sulphuric acid, but this is surely going a little too strong for sulphuric acid dissolves both silver and copper and is more apt to injure than benefit the plates.

The substances deleterious to amalgamation are grease, which may be neutralized by a mineral soap, lye, caustic potash or soda, cyanide of potash, etc., finely divided iron, or base metallic oxides, with cyanide of potash, free acid or acid sulphates with lime, antimony and copper salts with common salt. Some amalgamators use hydrochloric acid for cleaning plates coated with zinc, manganese or copper sulphate, but as a general rule, acids should never be used on silvered plates, although in this case the acid does not dissolve silver.

Often the ore is so acid the verdigris appears in spite of all precautions. When these spots increase to the proportion where amalgamation is seriously interfered with the plates should be replated.

In mills where plain copper plates are used and the amalgam allowed to remain on the plate until clean-up day the procedure is different than in the usual California mill. It may be necessary to steam these plates previous to scraping off with a chisel. The table is covered with a wooden cover leaving a space between it and the copper for steam which is introduced through a hole in the top, under say 50 lb. pressure. The steam is allowed to act for twenty minutes and is turned off. The cover is removed and the amalgam which is then the consistency of cheese is scraped off with chisels. When the amalgam has been scraped off, the plate is dressed with weak cyanide of potash, fresh mercury is rubbed in and spread evenly over the surface. Scaling plates is a more drastic method of cleaning applied to old and discarded plates. Sal ammoniac, nitre and hydrochloric acid in equal proportions diluted with water are applied to the plate with a soft brush. It is allowed to stand twenty minutes and then heated over a fire. When quite black it is dipped into a water bath, when the scaling can be washed off. The mercury may be driven off before applying the mixture.

The illustration Fig. 17 shows the position of plates and amalgamating surfaces in a mill where amalgamation is the important part of the process of gold recovery. Inside the mortar we have

5

the back and chuck-block plates, directly outside the mortar
we have the splash and lip plates, then we have the long table
plate which may be either in one piece or in 4 ft. lengths with a
step down between each length.   One does not now see the narrow amalgamating plates which came after the short wide
apron plate.   The table plate is now usually the full width of the
mortar.   It is a good plan to put four or five riffles at the end

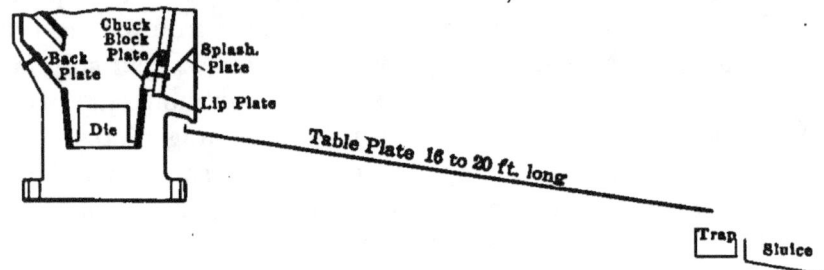

FIG. 17.—Position of amalgamating plates.

of the plates just above the end box.   Next we have the mercury
trap and lastly the sluice that conveys the pulp to the cyanide
annex or to waste.   Old shoes and dies may be placed in the
sluice.   They make excellent riffles and it is surprising what may
be cleaned up from them after a year's run.

The amount of amalgamating surfaces will be about

Back plate	5×54 in.	270 sq. in.
Front plate	5×50 in.	250 sq. in.
Splash plate	12×50 in.	600 sq. in.
Lip plate	8×50 in.	400 sq. in.
Table plate	60×240 in.	14,460 sq. in.
Total		15,920 sq. in.

If the battery crushes four tons per stamp per day this will give
786 sq. in. or 5.4 sq. ft. per ton per day.

Some engineers figure the amount of amalgamating surface by
the number of stamps, but this does not appear reasonable for
some stamps crush 10 tons a day while others crush only 2 tons.
It is not reasonable to suppose that both of these stamps require
the plate area when five times as much material is being reduced
by one than by the other.   In the South African Mills the plate
area varies from 9 to 15 sq. ft. per stamp.   These stamps have
a high duty with a coarse screen so that the conditions do not
apply to the gold quartz ore of the United States where the

practice of catching only that gold not easily cyanided has not come into vague. We aim to catch all the gold possible on the plates and should a cyanide plant follow the amalgamating tables we still aim at high amalgamation. The present aim in South Africa appears to be to reduce the plate area, taking them away from the battery altogether as has been done in some

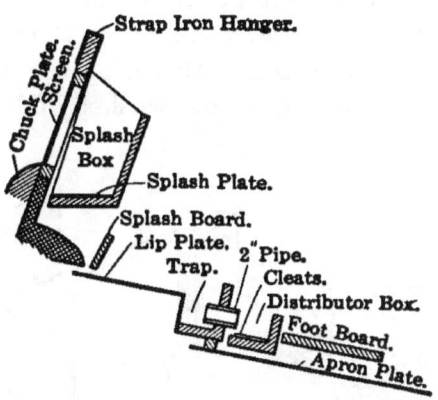

FIG. 18.—Amalgamating plates.

mills in the United States, but putting them altogether in the tube mill circuit; that is running, the sands through classifiers and tube mills before amalgamating on stationary tables.

Fig. 18 shows a section of the Laurentian Mill where the arrangement of plates is somewhat different than that of the preceding illustration.

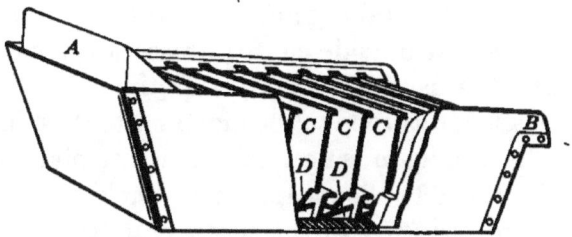

FIG. 19.—Pierce amalgamator or trap.

Mercury traps are put at the end of the plates to catch any stray pieces of amalgam or drops of mercury that may escape from the plates. The presence of amalgam or mercury in the trap does not necessarily indicate poor amalgamating, but it does indicate the necessity of a trap. Several different traps are illustrated, the Pierce Amalgamator, Fig. 19, being the best.

Fig. 20 is the usual type, while Fig. 21 is that recommended by the Colorado Iron Works.

The monthly clean-up is generally conducted as follows: The battery is thoroughly stamped out until the stamps begin to pound, the water is now turned off just before the stamps are hung up to expel all the water. The plates are now well wetted with mercury to hold any loose amalgam. The screen is now taken out and a bucket put at the end of the plate under the exit pipe to catch all the sand going down the plate. The inside plates are now taken out and with the splash and lip plates are laid on the table. These are now washed off, while another set

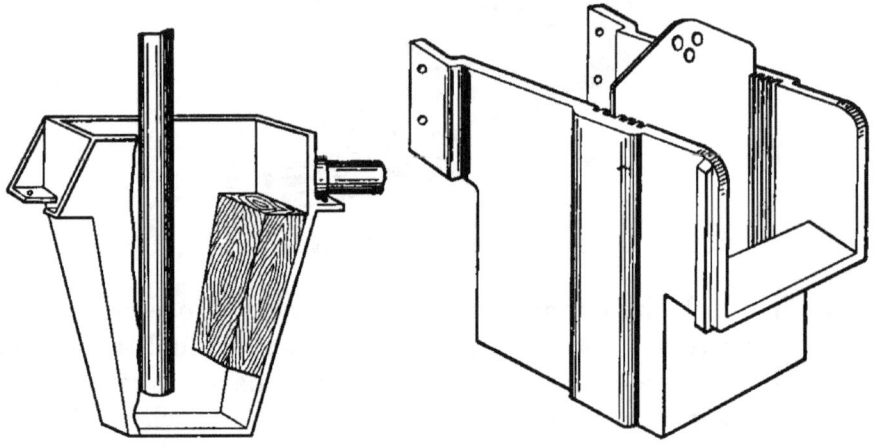

FIG. 20.—Mercury trap.   FIG. 21.—Square form of trap.

of men dig out the dies, taking all the sand to the amalgam barrel. The hard inside and outside plates are scraped with a chisel, a bent file being the usual tool. By this time the dies have been taken out, cleaned, returned to the mortar after turning to equalize the wear. The amalgam is now taken off the plates and they are treated as in the day clean-up. The back plates and chuck-block plates are returned to the mortar, meanwhile the tappets have been reset if necessary for the dies may have keen worn out and a new set required, or the shoes may have needed changing. The screen is now put on, ore is raked down into the mortar, water is turned on and the battery dropped. The sand from the end of the plate goes to the amalgam barrel which is treated to a can of lye and sufficient water to form a thick paste and is re-volved for twenty-four hours.

Five-stamp batteries may be cleaned at the rate of one per hour with sufficient help. It is a saving of money to take men out of the mine on clean-up day to help in the mill, for time saved there is important. In a small mill the amalgamator usually does all the work and it may take him a whole day to clean a five-stamp battery.

Each five-stamp battery requires from 1/8 to 16 oz. of mercury per twenty-four hours. The loss of mercury is due to the character of crushing, the constituents of the ore and the purity of the mercury before using. The loss may vary from 1/100 of an ounce where amalgamation is practised on the outside alone to an ounce a ton in exceptional cases, the average being from 6 to 10 dwt. per ton of ore crushed.

That copper plates absorb gold may influence their amalgamating ability, at all events this absorption takes place to saturation. The limit is placed at about an ounce to every 5.4 sq. ft. of surface, one-eighth of an inch thick. The effect of silver-plating is to retard the rate of absorption. The rate roughly follows the grade of the ore. That old plates well saturated amalgamate better than new plates is well known. This emphasizes a previous statement that amalgam catches gold better than mercury alone.

It may appear to some that due consideration has not been given to the use of the electric current in amalgamating gold ores. The reason is plain—millmen are seldom electro-chemists and few mills can afford the luxury of this extra man. The amalgamator may go so far as to lay strips of iron on his plates or to connect the iron work of his battery with the poles of a generator, but to be able to follow the complicated chemical changes due to the electrolysis of water and other chemicals added, and in the ore would take special training. There are special cases where the electro-current might be of benefit and cases where it would be harmful. It is a question of experiment, not of sentiment. This class of amalgamating has come into disrepute because the charlatan has too often used electricity to catch the unwary. Under electric action copper rapidly disintegrates, but iron can be made to amalgamate, as with the use of sodium amalgam. This limits us to the use of a separate receptacle or the use of a chemical that will protect the copper and at the same time enliven the mercury. The first condition is obtained with mercury baths or wells kept in electric tension.

the latter by the use of a chemical which will set free the chlorine constituent.

The electrolytic process of amalgamating may be classed under three heads: First, forming an electric couple with the amalgamated plate by means of strips of iron on the plates or using a composition plate composed of two metals. Second, mercury sluices or baths where a sheet of mercury is kept under electric tension, the electrolyte being usually a solution of salt or cyanide of potash. Such for example is the Molloy Pan, where mercury is separated from the carbon or lead anode by a sheet of porous clay. Third, where mechanical stirrers are used such as in the Palatin and Clerici vat, the bottom of the vat containing mercury and copper plate cathodes, the electrolyte being cyanide of potash and salt.

Bearing a close relation to the above electrolytic methods are those using sodium amalgam either by adding a prepared amalgam of sodium and mercury on the plates, or in a separate apparatus preparing the sodium amalgam out of contact with air or water and circulating the gold-bearing pulp over such prepared mercury.

The sodium amalgam liberates hydrogen gas and may deoxidize oxides, but whether beneficial or not will depend upon the constituents of the ore. Usually those ores not easily amalgamated are those that would suffer by any free hydrogen, that is, the acid ores which contain soluble sulphates. Sulphuretted hydrogen would form, which would have the effect of fouling the mercury.

Most electric processes use sodium chloride, assuming that an electric current will set chlorine free. This is a fallacy, for chlorine cannot be free from salt by an electric current unless the salt is perfectly dry. Mercuric chloride is another salt used by electro-chemists to promote amalgamation, the theory being that chlorine is set free and combines with the common base metals, while the hydrogen set free from the water reduces the hydrates.

In all the literature on milling there are very few recorded instances where the use of electricity has been beneficial in amalgamating gold ores. The following case is worthy of notice:

"The mortars were provided with interior amalgamating-plates in communication with the poles of a dynamo that produced a current of 150 amperes, 14 volts. The two-stamp batteries discharged into a common channel, in which, side by side, were placed the large amalgamating plates, one communicating with the positive pole, the other with the negative" . . . "the electro-motive power was then, not in series, but in tension. This produced the desired result. The liberation of gases diminished considerably, and the loss of mercury became insignificant."[1]

[1] *Trans.* Am. Inst. Min. Eng., Vol. XXXII, page 487.

CHAPTER VII

STAMP MILL CONSTRUCTION

The first requisite in the construction of a stamp mill is a good site where there will be sufficient grade to ensure a process where gravity will assist the passage of the ore from stage to stage. It must be accessible for fuel and water and the ground must be solid for laying good foundations so that there will be great stability, or lack of movement and vibration. The mortar blocks should be devoid of movement and the ore bins solid and well anchored. There should be plenty of room in the mill and nothing crowded. To obtain these necessary conditions the machinery must be first class, the construction good and the plan of the mill well thought out beforehand with all the necessary data to form a proper selection of machinery best suited to the work to be performed.

The successful operation of a stamp mill, especially where the margin of profit is small, is greatly influenced by close attention to details.

It was customary in days gone by to run twenty stamps from a single cam-shaft. This required the whole twenty to be hung up if the belt needed relacing, or a stem had to be changed. It is usual now to have ten stamps on a cam-shaft with the bull-wheel on one end or in the center of the shaft. But for the vibration a separate shaft for every five stamps would be the best construction.

After one has decided this point—the number of stamps in the mill and the number of stamps on a cam-shaft and the position of the bull-wheel he must decide the following details which will be treated in the same order. First the form of battery construction, that is the shape of the frame work, second the form of mortar foundation, third the type of battery post foundation and then such details as the shape of the mortar, size and material

72

for cams, cam-shafts, bull-wheels, stems, shoes, dies, boss-heads, cam-shaft boxes, tappets, guides and the type of ore feeder.

Battery Frames.

There is little variation in the design of battery frames being erected at the present time. They are generally of the back-knee pattern, the battery posts being bolted to the ore bin sill timbers. Figs. 22 and 23 show the timber work of this style of framing. An older style and one still preferred by some millmen is the front-knee pattern. This has the disadvantage of cutting out light

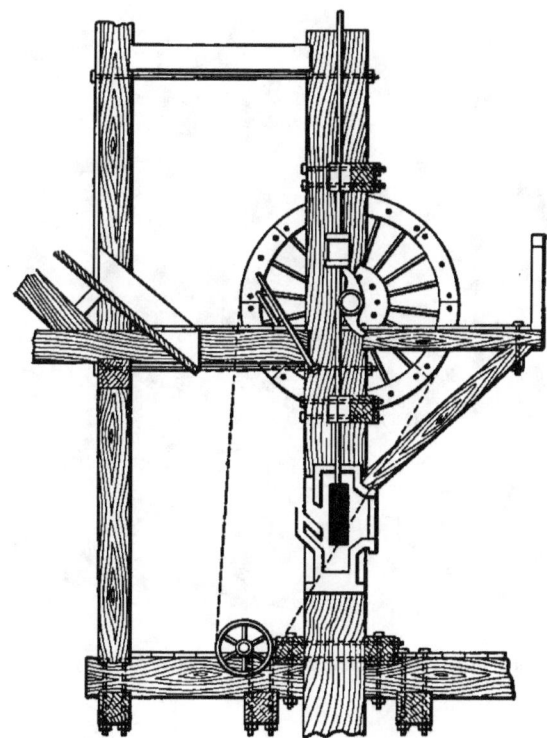

Fig. 22.—Back knee frame.

from the battery plates and has proven weak with heavy stamps requiring a back-knee brace to give stability. The only point in its favor is that the tight side of the driving belt is uppermost, but the same end can be attained with the back-knee by having the driving pulley on the ore bin foundations. Fig. 24 shows the timbering for this style of construction. The "A" frame is one seldom seen and is here illustrated, Fig. 25. The double-post frame as typified by the Hendy single-unit batteries is illustrated in Fig. 26. Both the Hendy and the Nissen form of construction for units of three or more mortars use a suspended middle battery post called the "jack post." This is the weak part in the construction. It increases the vibration in the frame and is a continual source of trouble. This post should be extended to the concrete foundation. It would increase the

FIG. 23.—Timbering for back knee frame.

floor space somewhat, but would diminish the vibration of the structure. The heavier the stamps the more vital is this defect.

A ten-stamp mill may be either constructed with three posts or with four; in the former the bull-wheel is at one end, in the latter the pulley is in the center between the two center posts. This latter style of construction has proved successful and distributes the strain on the cam-shaft and causes less vibration than the three-post frame. The two variations are illustrated in Figs. 27 and 28.

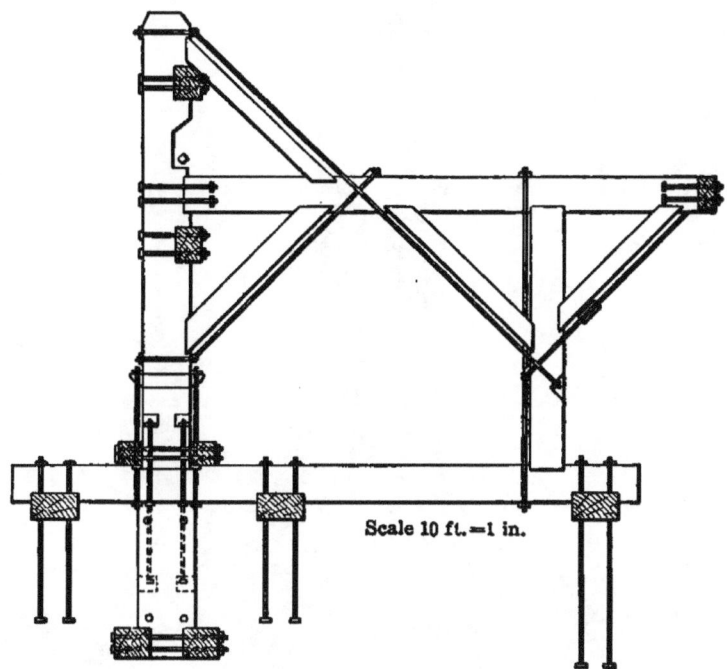

Scale 10 ft. = 1 in.

FIG. 24.—Front knee frame.

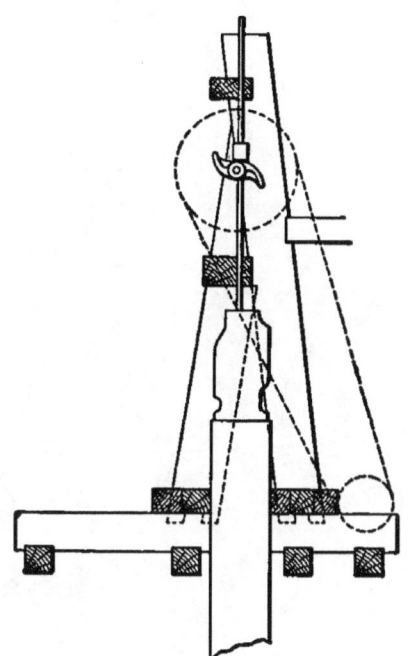

FIG. 25.—"A" battery frame.

FIG. 26.—Two post battery frame.

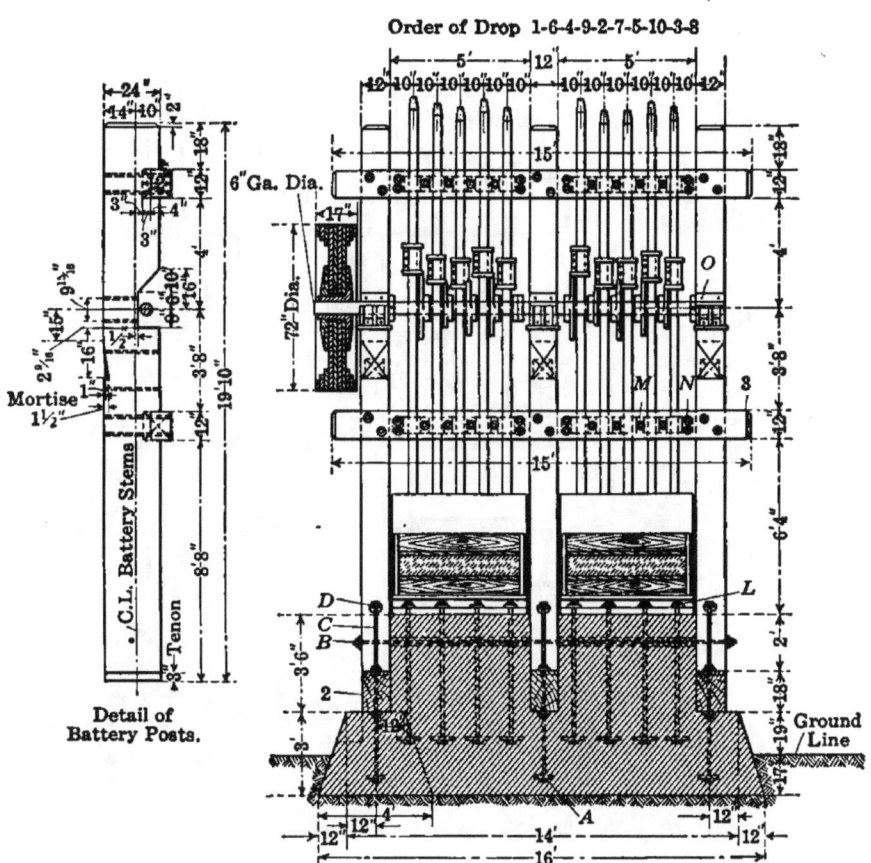

FIG. 27.—Three post ten-stamp mill.

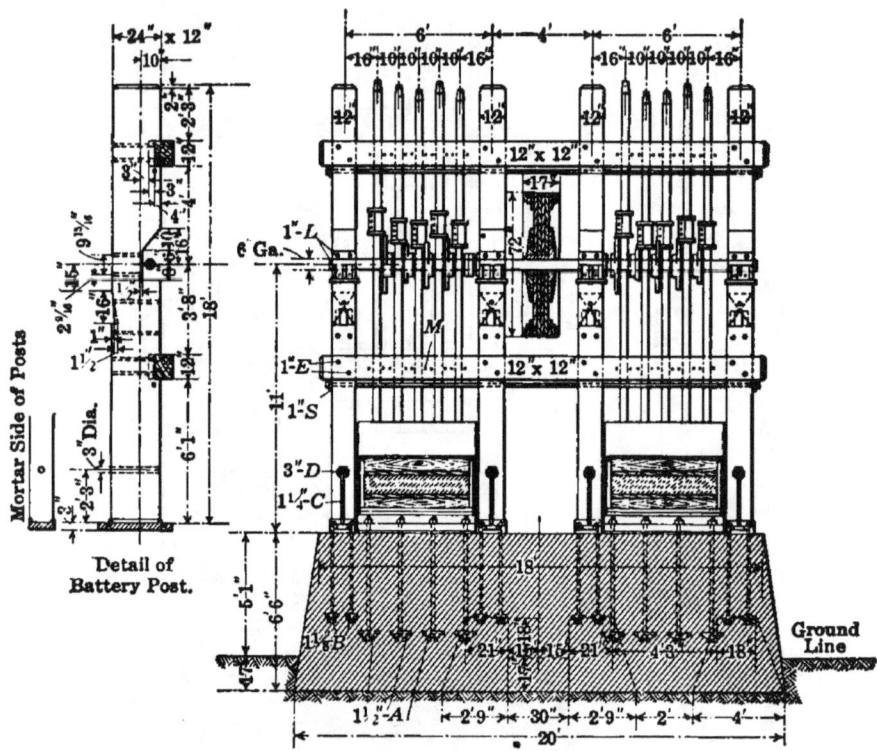

Fig. 28.—Four post ten-stamp unit.

Mortar Foundations

Mortar blocks are made either of wood or concrete, the latter being the material of which most of all the late mortar blocks have been made. In every way they are superior to the wooden

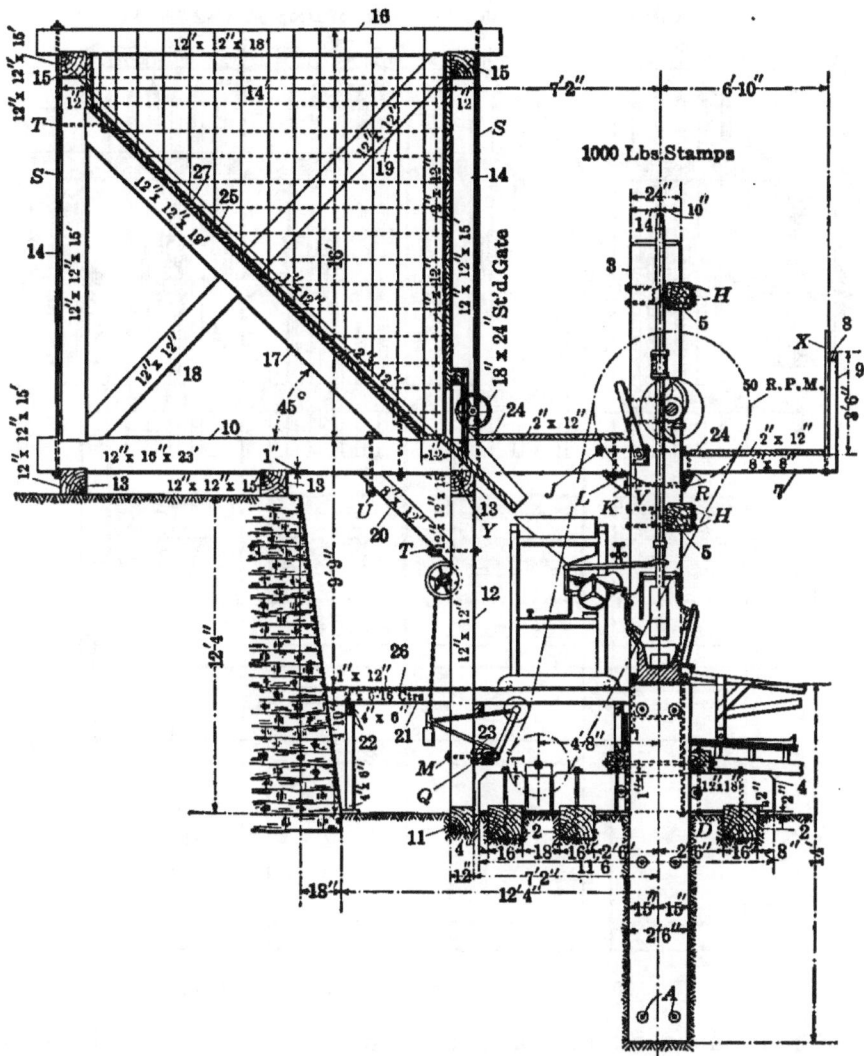

Fig. 29.—Wooden mortar block construction.

blocks, but as there are some who still believe in the wood construction a short description may not be out of place.

Wooden mortar blocks may be made of timber of any size from pieces the size of the base of the mortar to that of a 2-in. plank; when of large dimensions they are bolted together unless of one piece, and when of planks these are nailed and bolted together, the planks being put on lengthwise of the mortar, so that if the outside planks rot they may be stripped off and new ones nailed on.   If put on parallel with the smaller dimension of the mortar this cannot be done without taking out the battery posts.

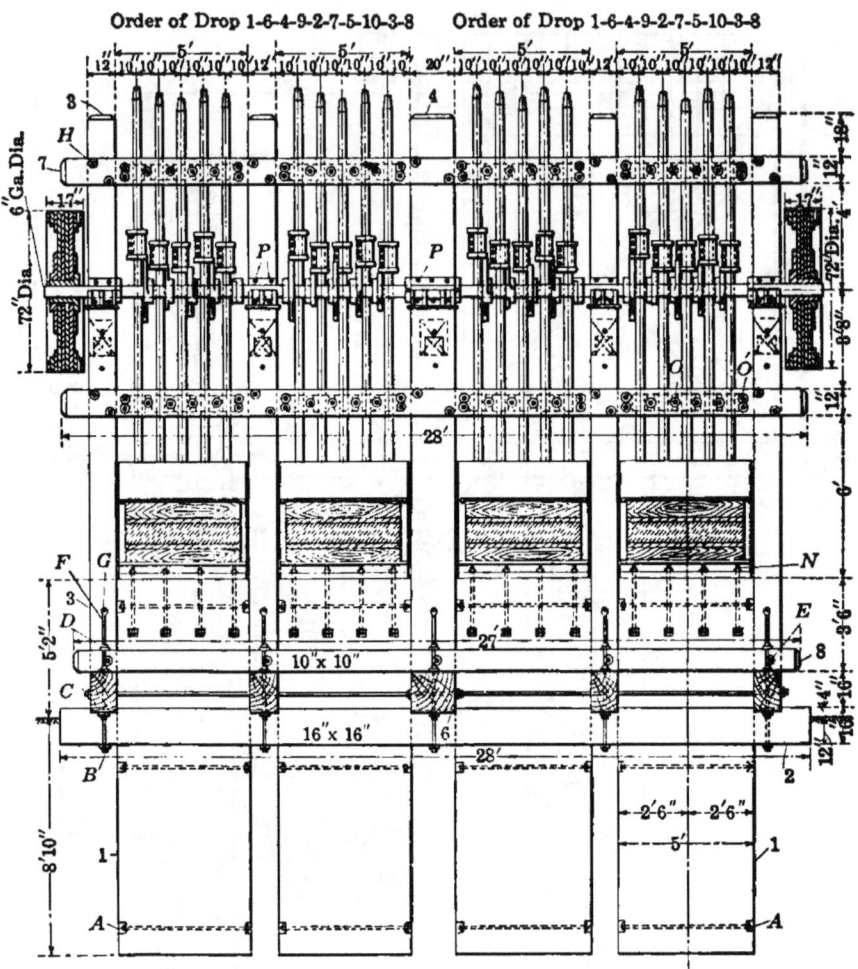

Fig. 30.—Wooden mortar block twenty-stamp mill.

For the mortar block, a hole is dug of the right depth, preferably in solid bedrock and wide enough to leave about 2 in. clearance around it, which latter is filled with concrete or tailings from the battery according to the preference of the millwright. These mortar blocks vary from 8 to 15 ft. in length, with 6 to 9 ft. in the ground.   It is evident that the deeper the mortar block is in the hole, the more stable will the block be.   The block may lie on a concrete base or on a double layer of planks placed crosswise and spiked to each other.   This block should be made of spruce or sugar pine, for these trees resist dampness better

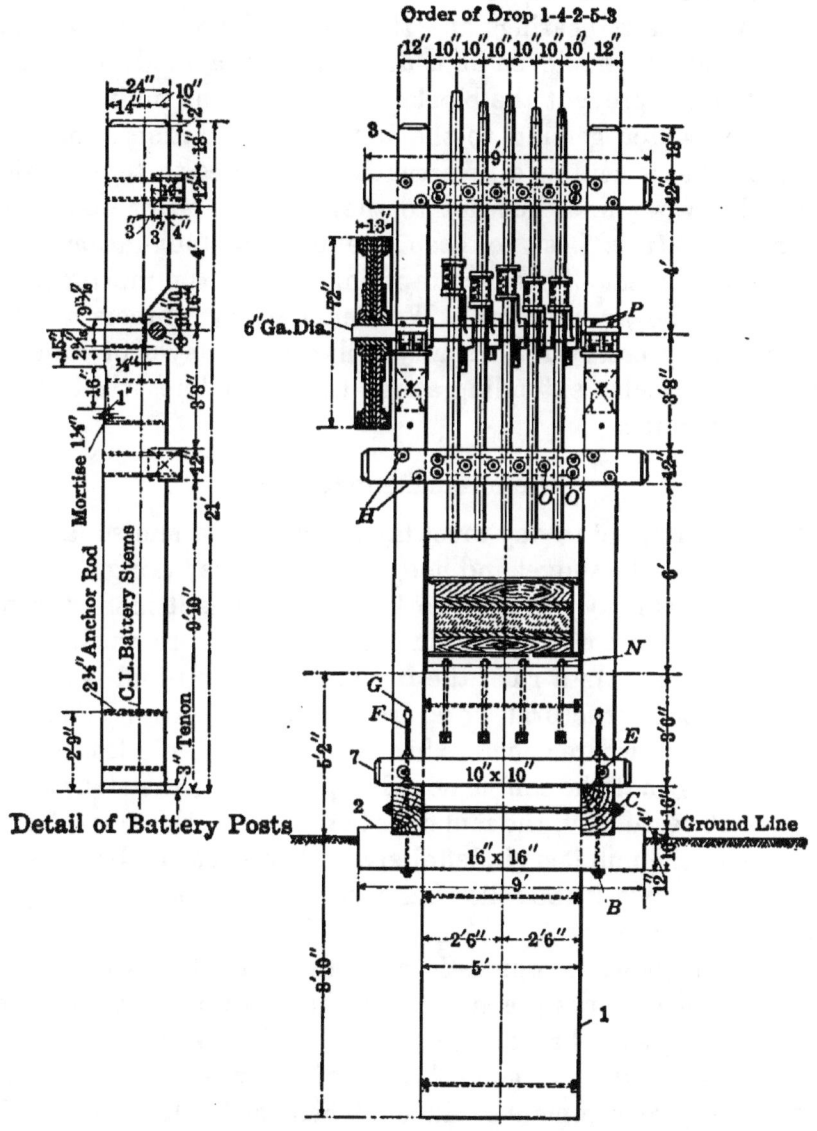

Fig. 31.—Wooden mortar block five-stamp unit.

than other classes of pines.   The top of the mortar block should be planed true and covered with a sheet of rubber a quarter of inch thick to make a good joint with the cast iron mortar. Fig. 1 shows this form of construction and Figs. 29, 30 and 31

6

show the Hendy Standard form of wooden mortar block construction.

The "cellar" under the feeder and amalgamating floors should be high enough for a man to walk in comfortably so that mortar blocks, line-shafts bearings and the foundations generally could be inspected. It should be kept dry and with a good draught of air. This will prevent the blocks rotting. Some millmen favor the idea of water getting to the battery foundations, for they claim the water acts as a preservative. It does to that portion below the water line, but not to that portion on or above the water line. It is best to keep the battery foundations dry, even to the extent of a concrete filling around the mortar block and above the ground line. Wooden battery timbers treated with a good preservative, an airy cellar and a layer of concrete next to the blocks, extending above the water line, will last the life of most mines.

Concrete Foundations

The millwright of to-day must lay aside his saw and square and take kindly to the shovel and hoe, for concrete is taking the place of much of his heavy timber work. Not only is it more durable than timber, more economical in cost per ton of ore crushed, but it looks better. As a rule, the inside of a stamp-mill is a dirty place, and the structural work crude and unsightly. Compare this with a mill where concrete has been used to its fullest extent. We find now that the mill is neat in appearance, the structure is better architecturally, the mill does better work, the extraction is better, and the mill lasts indefinitely. To obtain the best results from concrete work the following observations and rules may be studied with interest.

The best grades of domestic cement should be used. While some foreign cements are superior to our ordinary brands, even for cementing silex brick in tube mills, a class of work requiring the best materials, a good brand of domestic cement will answer every requirement. Except when used in large quantities particularly where the cost of labor is high, it is better to hand-mix all concrete. The mixture of sand and rock, which with cement makes concrete, is often called the agglomerate. Concrete should always be mixed fresh; that which has been mixed over half an hour has had an initial set and should be rejected. The sand should preferably be clean and sharp, although rounded sand-grains or sand that contains from 10 to 15 per cent. of dirt has been found not to lessen the strength of the set. The broken rock used should not be larger than a size which will pass a 3-in. ring, and should be well washed before mixing to remove all dust and dirt, so that the cement and sand mortar will adhere to the surfaces.

Cement will not readily take up any more water than is necessary for mixing; the excess will come to the top, and does no harm unless the excess is carried to an extreme, causing a separation of the cement from the agglomerate, thereby weakening the mixture. Wet concrete takes longer to set and eventually becomes harder than when the water is only in sufficient quantity to create adhesion. Concrete of this latter description should be well rammed until water comes to the surface. While setting, concrete should be kept wet, for this retards the setting, and as stated above in respect to wet concrete, it produces a more lasting structure. It is not necessary for any piece of work to be of the same proportional composition throughout, for the richness of a concrete should be proportional to the strains it must bear; for example, the base of a foundation, usually being larger than the top, may be made of a poorer mixture and yet stand the same strain per square inch as the top.

Concrete mixed with fresh water should not be allowed to set when the temperature is below freezing, unless there is a heavy weight upon the structure to keep the mixture in compression, or unless fires are kept burning near the structure, otherwise the concrete may be loose and crumbly. In a recent reference to this subject it is claimed that freezing will not materially injure concrete, for while frozen the setting action ceases to start again when the water thaws. When the temperature is below freezing, the usual plan is to add some chemical to the water that will lower the freezing-point. Salt is most commonly used, and when not in excess of 10 per cent. of the weight of the water, it makes a stronger structure than fresh water. The salt delays the setting, and lowers the temperature at which the water will freeze. There have been various rules proposed to determine the quantity of salt necessary to prevent freezing. "Add 1 per cent. by weight of salt to the weight of water for each degree F. below freezing," or "1 lb. of salt to 18 gal. of water for a temperature of 32° F., and an increase of 1 oz. for each degree lower."

Concrete may be mixed with hot water, and sets more rapidly the hotter the water. A temperature of 150° F. is a safe limit, unless the mass is immediately transferred to the form. It is often necessary in cold weather to heat the rock and sand which may have frozen over night. It is best in this case to thaw the sand before a fire or in a drier, for wet sand and cement mix poorly. In case it is not practicable to use a drier, then the sand should be spread out, the lumps broken up, mixed with cement, and hot water used to make a mortar. This will require more work than when dry sand is used, but if time be taken, a good mortar may be made. The rock may be drenched with boiling water.

In structures where the strains are chiefly compressive, the surface of the concrete laid on the previous day should be cleaned and wetted; no other precaution is necessary. If the concrete has set, or it is necessary to lay it intermittently, the surface should be roughened or left with corrugations. Before adding the next batch of concrete, the surface should be cleaned and wetted. Concrete is mixed by volume, not by weight, and all calculations are made upon this basis, although the millwright will generally prefer the bag of cement as his unit and the number of tons of sand and rock. The conversion of volume to tons is simple. The next question is, what should the composition be for a given piece of work? The theory is that, for concrete to be at its highest efficiency, the spaces between the sand should be filled with cement, and those between the rock with cement and sand mortar; therefore the amount of voids or empty spaces between the particles of sand and rock will govern the proportion of the ingredients.

For extensive or accurate work cements are always tested, both chemically and physically, including the percentage of voids in the sand and rock it is proposed to use. The millwright, as a rule, will not bother with these refinements, and will use some arbitrary measure that has proved successful in other similar work. We will consider an average sand and an average rock, and endeavor to work out our problem, first explaining the mode of mixing the constituents of concrete as found by experience to give the best results. This is a concrete engine-foundation specification for one of our big railroads, and is a fair example of what is considered necessary when mixing concrete.

" About half the sand shall be spread evenly over the bed of the mortar-box, the cement shall be spread evenly over the top of the sand, and finally the remainder of the sand shall be spread on top. The sand and cement shall be thoroughly mixed by turning and re-turning with a shovel. The mixture shall be drawn to one end of the box, water poured in at the other end, and the mixture drawn down to the water with a hoe, a small quantity at a time, and mixed vigorously until there is a stiff mortar. The mixture shall then be leveled off and the required amount of broken stone shall be thoroughly wetted and spread over it. The whole mass shall then be thoroughly mixed by turning and re-turning with shovels into rows, at all times preserving the same thickness of the mass until the mortar completely fills all the interstices of the rock. After transferring to the forms, ram with 20-lb. hammers until water comes to the surface."

The requirements for good concrete are, then, thorough mixture of dry sand and cement, the addition of water to form a mortar and the thorough mixture of broken stone with this mortar.

To make a concrete where all the voids in both sand and rock are filled, and using the figures for sand and rock as above, the mixture should be 1 cement, 2.5 sand, and 4.6 rock. This is a mixture in which all the spaces between the sand and rock are filled. Experience teaches us that a better result is obtained when the rocks are separated by a layer of mortar and the grains of sand by a layer of cement. This 1: 2.5 : 4.6 mixture does not allow for mistakes in testing or leave allowance for variations in sizes of the material used. By actual tests the best mixture for concrete has been found to be nearer 1:1.5:2. This mixture should bear compressive strains up to 2900 lb. per square inch. When the weight on a foundation is excessive, or subject to sudden changes, the concrete must be at its highest efficiency; therefore a rich mixture such as 1:1.5:2 must be used, but the less the strain on the foundation the poorer the mixture may be, leaving always a safe allowance for the unforeseen, which always happens. Good concrete made with a mixture somewhat richer than that found by calculation will stand a strain of from 2100 to 2800 lb. per square inch. Under tension, concrete will stand from a sixth to a tenth of this, or 200 to 300 lb. A stamp-mill foundation figures out thus:

The base of a mortar for concrete foundation is usually 32 by 60 in., or 1920 sq. in.; the weight of mortar, stamps, dies, ore, etc., about 14,000 lb., or about 7 lb. to the square inch. Two battery-posts 12 by 28 by 21 ft. with cam-shaft, bull-wheel, cams, etc., will weigh about 14,000 lb. The base with sole-plates occupies about 720 sq. in., giving 20 lb. to the square inch. The compressive strain on the foundation of the battery-posts is therefore over twice that on the mortar base. Owing to the vibration caused by the dropping and raising of the stamps, the tension on the foundation bolts will more than equal the compressive strains, therefore the foundation must be at least ten times as strong as if the compressive strain were alone considered. The following table shows the compressive strains that certain mixtures of good concrete will bear:

Cement	Sand	Rock	Pounds per square inch
1	1 1/2	2	2800 to 2900
1	2	3	2500
1	2	4	2300 to 2400
1	3	4	2100
1	3	5	2000
1	4	6	1700

It would appear from this that a 1:4:6 mixture, if made with good materials, would leave a big factor of safety. The mixture generally used for battery work is from 1:4:4 to 1:3.5:3.5. To bring this up to 1:4:6 and still have a big factor of safety, large rocks up to 8 in. in diameter may be hammered or punched down in the center of the form as the concrete is built up. This makes a rubble concrete and will not materially lessen the strength of the foundations. Suppose that now we have determined upon a certain mixture, we may want to know the quantity of materials necessary. There is a simple rule which will give approximately the number of barrels of cement in a cubic yard of concrete. Divide 10.5 by the sum of the parts of all the ingredients. For example, in a 1:2:3 mixture there will be 10.5 divided by 6, or 1.7 bbl. per cubic yard of concrete. If the work is of great magnitude, the proportions may be found by actual tests. Knowing the number of cubic yards of concrete and the proportions, and figuring 22 cu. ft. of sand and 20 cu. ft. of broken rock to the ton, the weights of the ingredients may easily be found.

For example, the mortar and battery-post foundations of a ten-stamp mill will contain 20.5 cu. yd. Suppose we use a 1:4:4 mixture; then 10.5 divided by the sum of the ingredients, 9, will give us 1.15 bbl. of cement per cubic yard, or 20.5 by 1.15, or

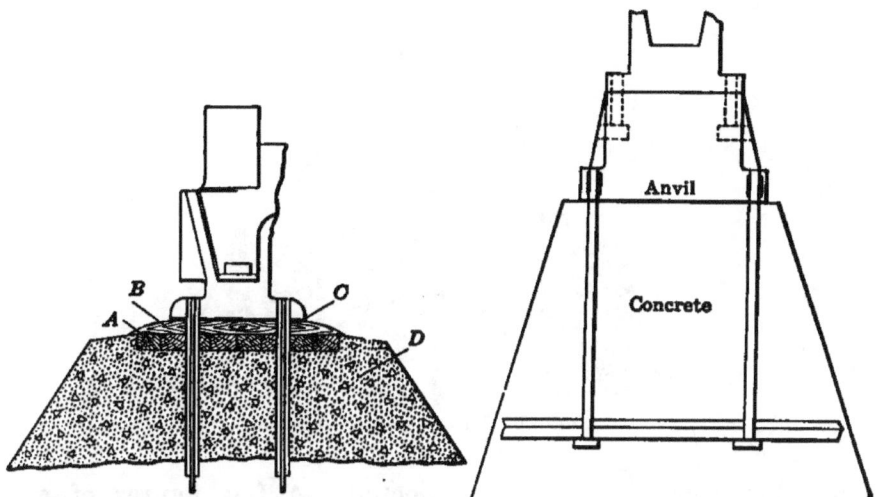

FIG. 32.—Silver peak. Nevada construction. *A*, *B*, Maple base laid in tar. *C*, Rubber blanket. *D*, Reinforced concrete.

FIG. 33.—Anvil block foundation.

23.5 bbl. of cement, or 94 bags of Portland cement. The cement and sand being as 1 to 4, therefore, as a barral of loose cement occupies about 4 cu. ft., the sand will occupy 16 cu. ft. per barrel; 16 by 23.5, the number of barrels necessary, will give 376 cu. ft.

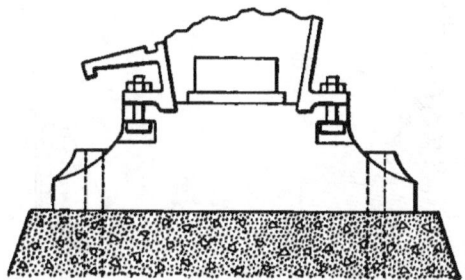

FIG. 34.—Heavy base with detachable upper section. (*Schmidt.*)

at 22 cu. ft. per ton, or 17 tons of sand. The rock and sand being in equal proportions, 17 multiplied by 22 divided by 20 will give us 18 tons of rock. As a barrel of cement weighs 400 lb. gross, the amount for this foundation will be 9400 lb., or about

8800 lb. net.   These figures are approximate, as we have assumed our data, which, though near an average, may not be correct for some specific example.   To make sure, the engineer had better allow an increase of 5 per cent. on each item for contingencies;

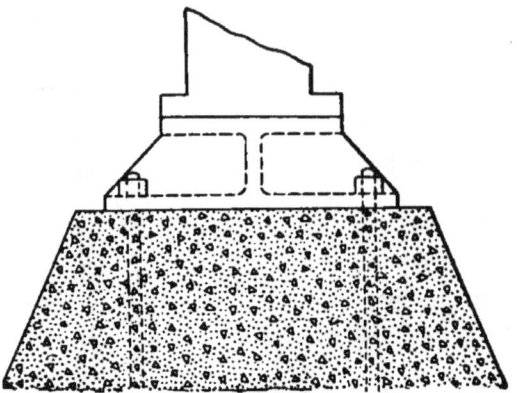

FIG. 35.—Vulture mill anvil block.

in other words, for the unexpected.   A few figures of what may be considered good foundations for water-tight work may be useful.   A 4-in. concrete wall will be water-tight under a head of 4 ft.; a 15-in. wall under a head of 20 ft.; while a wall 5 ft. thick will hold water under a head of 100 feet.

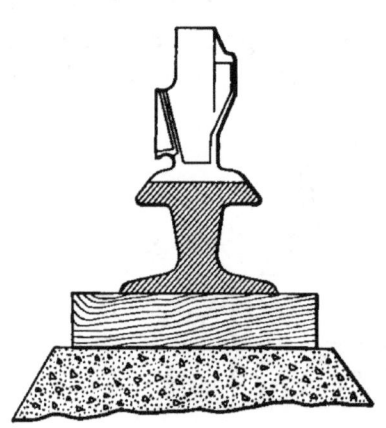
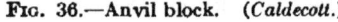

FIG. 36.—Anvil block.   (*Caldecott.*)

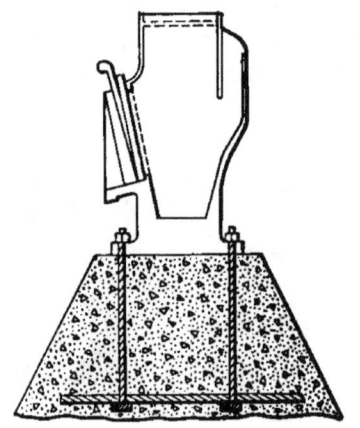

FIG. 37.—Anchoring by means of a bar.

In the application of concrete to stamp-mill construction there are two factors to be considered, first the weight the concrete must bear and second the liability to extreme vibrations.   The concrete is made rich enough to bear a weight many times that of the mortar and battery posts, but the vibrations must be reduced to a minimum by making the connection between the base of the mortar and top of the concrete as perfect as possible and by anchoring the mortar firmly on the concrete. When we come to

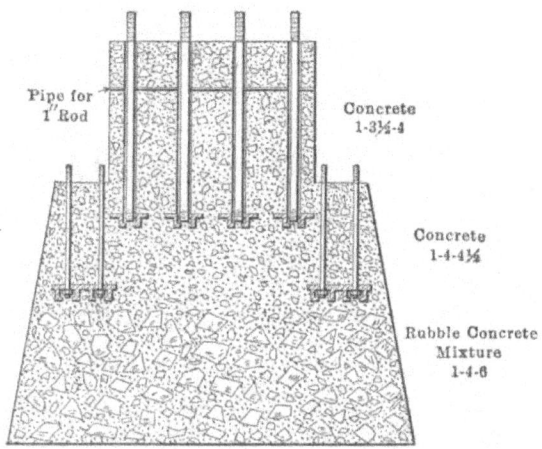

Pipe for 1" Rod

Concrete 1-3½-4

Concrete 1-4-4½

Rubble Concrete Mixture 1-4-6

FIG. 38.—Anchoring by means of cast iron washers.

consider battery post construction we will consider the feasibility of cushioning the vibrations of the cam-shaft which, however, does effect to some extent the vibrations we are now considering. The connection between concrete and base of mortar may be made

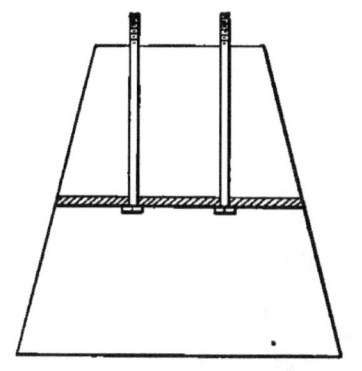
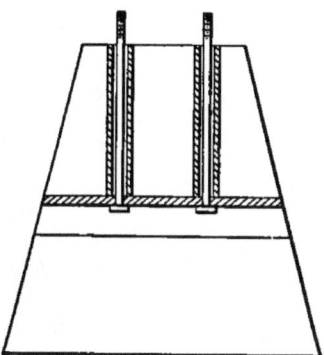

FIG. 39.—Anchoring with bar through concrete.

FIG. 40.—Showing bar and pipe around bolts.

with a sheet of rubber, a sheet of lead, by wooden blocks as at Silver Peak, Nevada, Fig. 32, or with an anvil block. The last is illustrated in Figs. 33, 34, 35, and 36. These are unnecessary and unsatisfactory. The best connection is made with a rubber sheet from an eighth to a quarter of an inch thick. This does not act as a cushion for it soon becomes as hard as iron, but it does help to make a close union. The mortar base is planed

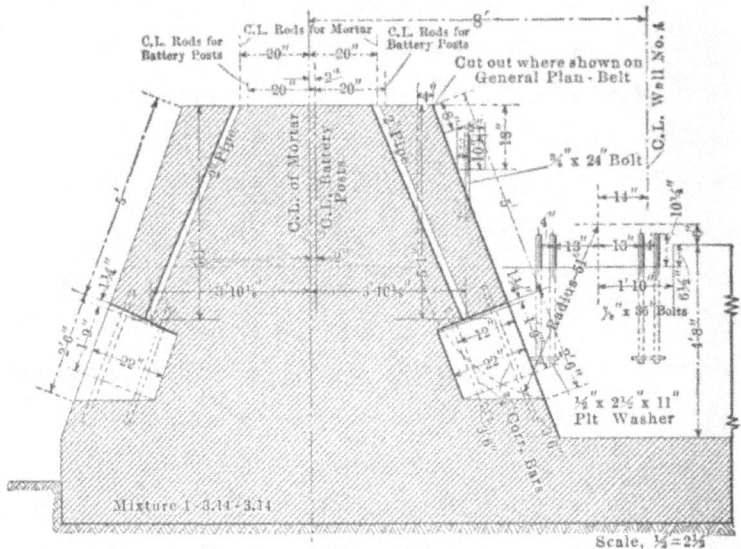

FIG. 41.—Goldfield Con. mortar block.

true in the shops, but if laid directly upon the concrete any movement of the mortar will wear into the concrete. The mor-

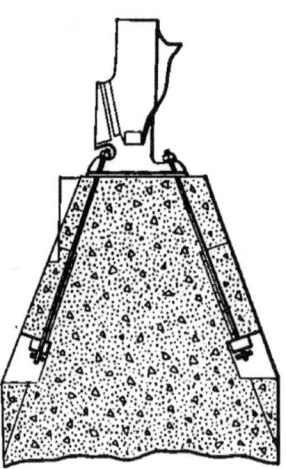

FIG. 42.—Mortar block 1670 lb. stamp. (*Coldecott.*)

tar is anchored to the concrete by eight bolts from 1 1/2 to 2 in. in diameter, preferably the latter.

The methods of anchoring these bolts are important and admit of many variations from the usual type. The lower ends may be held down by big iron washers or passed through a T-rail, two or more bolts being held by one rail. In these two methods the bolts are solid in the concrete and cannot be replaced. Figs. 37, 38, and 39 show this. Fig. 40 shows a hollow channel through the mortar block, with a T-rail anchor, and pipes surrounding the bolts.

The two best examples of the removable bolt type of anchoring are exemplified in the construction of the Goldfield Con.

Fig. 43.—City deep concrete mortar block.

Mill, Fig. 41, designed by Mr. Fleming and that of the 1670 lb. stamp in South Africa taken from Caldecott, Fig. 42. Fig. 43 shows the mortar block for the City Deep Mill, 2000 lb. stamps and is a good example of the removable bolt type of foundation.

The object in having a pipe somewhat larger and surrounding the hold-down bolts is not only to be able to pass the bolts through the pipe, but to allow a slight springing of the bolts in case they do not line exactly with the holes in the mortar.

The Cost of Concrete Foundations

The cost of concrete in place at the Goldfield Con. Mill is given as follows:

Cement	$5.03
Rock	1.06
Rock-sand	0.67
Pit-sand	0.45
Sand (total cost)	1.12
Labor	1.32
Forms	2.76
Superintendence	0.15
Reinforcing	0.38
Total cost per yard	$11.82

The cost of the concrete foundations for a ten stamp mill in Idaho is given as:

Labor, excavating	$32.30
Labor, foundation	68.30
Bolts and castings	50.00
Cement	33.70
Lumber	10.00
Gravel	30.00
Rock	7.25
Total	$231.55

This foundation contains 14 cu. yd. of concrete, but a safer and better foundation should contain 20 cu. yd. of concrete.

The mixture used for the Goldfield Con. Mill was 1–3.14–3.14, while that for the Idaho mine was stated as three sacks of cement to the cubic yard. I have used a 1–4–4 mixture with good results.

Battery Posts

Battery posts may be made of wood, concrete or iron. The first is generally used, the second is on trial at the City Deep, South Africa, the last has been often tried with doubtful success.

There are several methods of lessening the vibrations in the cam-shaft and battery posts. First, we have a bearing between each stem as in the Leschinger construction. Second, we may cover the cam-shaft boxes with a cap. This has been discarded as very unsatisfactory. Third, we may press the cam-shaft down with a block of wood under tension by means of a spring.

This has proven successful even at the cost of a little extra power. A proposed method is also illustrated, that of cushioning the vibrations by means of a heavy spring.

As important as the form of battery post is the method of anchoring it to the foundation. Fig. 44 shows seven different methods of anchoring a wooden battery post. The first two show the wooden battery post in connection with wooden mortar blocks, while the other five illustrations show this in connection

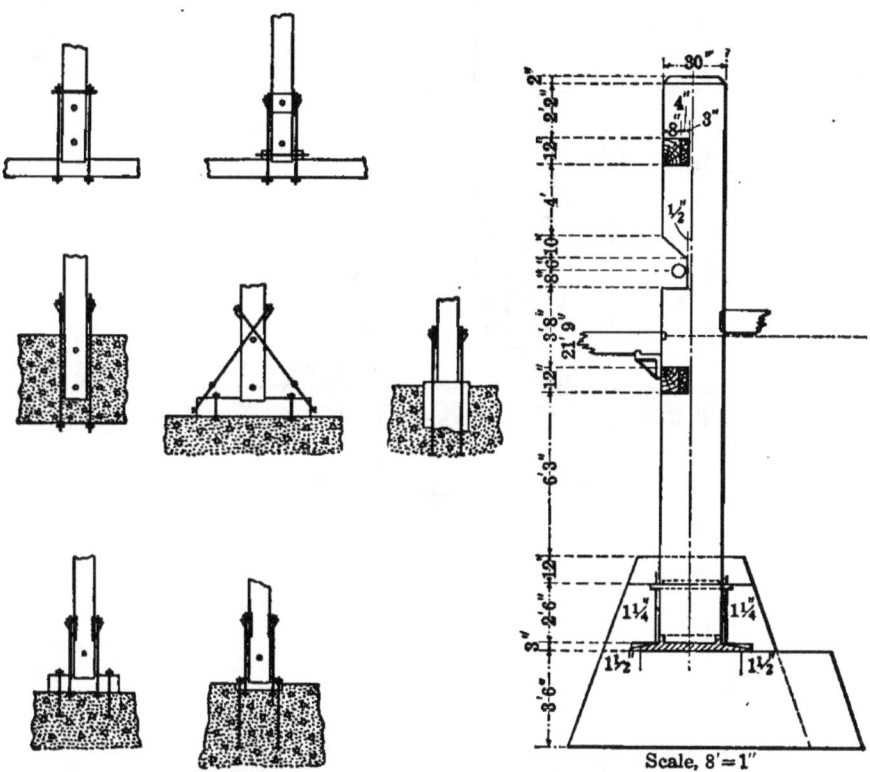

FIG. 44.—Wooden battery post anchoring.　FIG. 45.—Fort Bidwell Con. battery post.

with a concrete base; of the seven the last is by far the best construction, only instead of the brackets a 3 or 4 in. bar may be used in connection with the hold-down bolts to the post. This is shown in Fig. 45, being the battery post erected for the Fort Bidwell Con. Mines Co. Fig. 46 is the battery post at the Yellow Aster Mill. It is a poor form. There is no occasion for the short sill or for the hold-down bolts being at an angle; both give constant trouble. Fig. 47 shows a construction somewhat similar to the Fort Bidwell Mill, but that the plates are set away from the

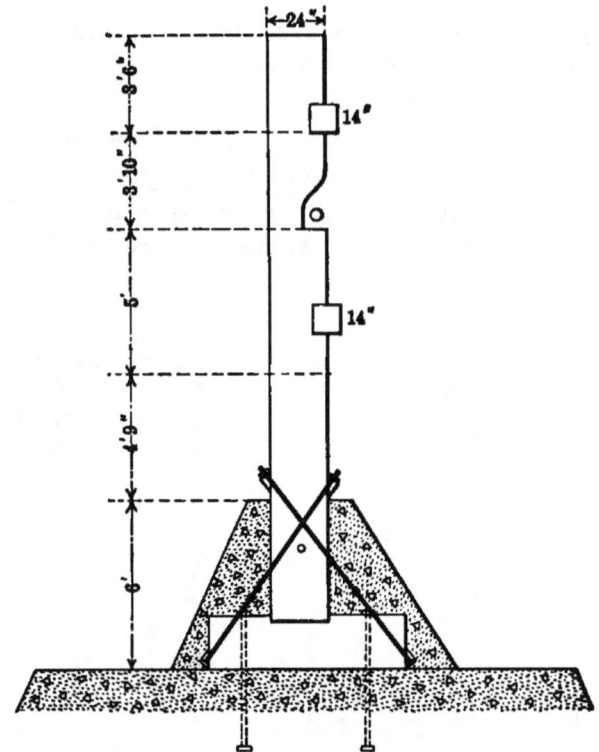

Fig. 46.—Battery post (Yellow Aster Mill).

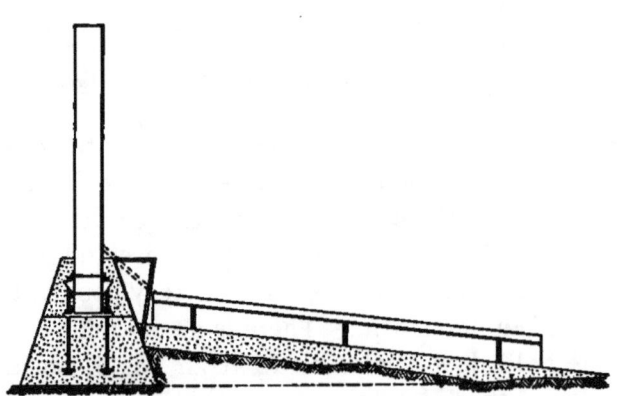

Fig. 47.—Showing plates away from battery.

battery, are stationary, leaving a space in front of the battery to run a car to carry shoes, dies and bosses.

Fig. 48 is a battery post of a South African Mill taken from Caldecott. The feature to be noticed in the independent support of the cam-shaft floor, a very excellent arrangement.

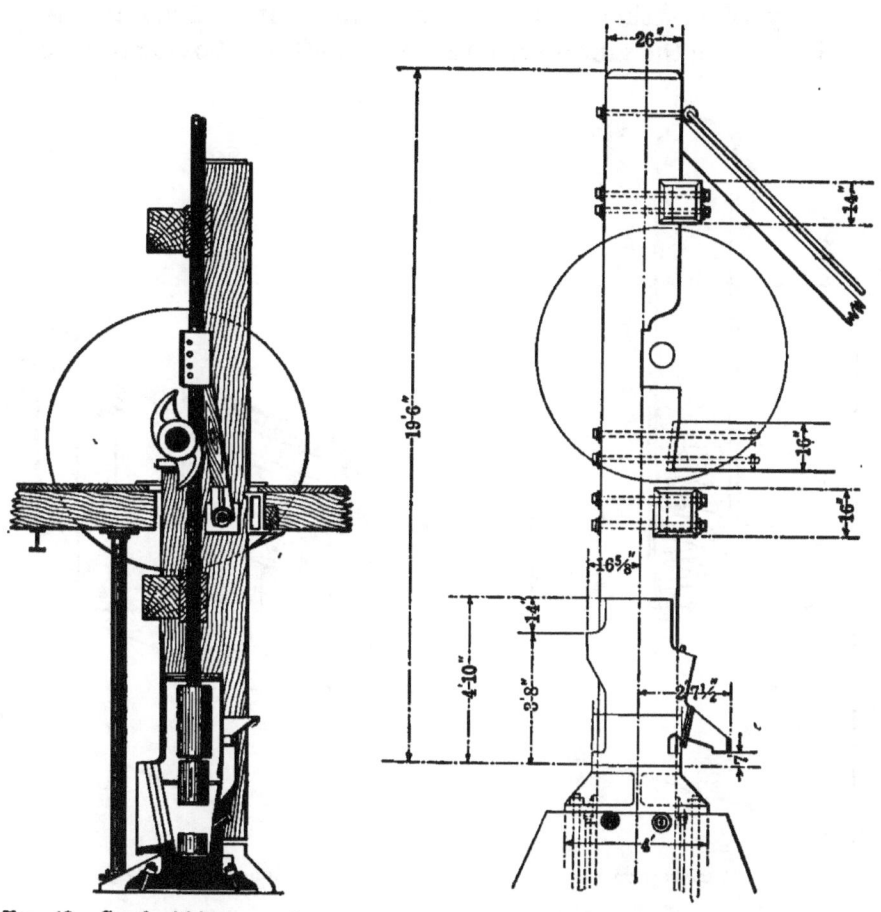

FIG. 48.—South African construc-   FIG. 49.—Elevation 1600-lb. stamp.-battery
tion. (*Caldecott.*)   (*D. D. Demarest Co.*)

Fig. 49 is the battery post of the 1600 lb. mill of the Vulture Mill, Arizona, erected by D. D. Damarest Co. This is a front knee frame, but I understand that it has been necessary to brace the battery posts to the ore-bin as in the back knee construction.

The objection to the all steel battery frame is the extreme rigidity. With the cam-shaft vibrations cushioned this objection would be negatived. There is little doubt that reinforced concrete is superior to steel in this connection. A feature of the reinforcement might be an 8-in. pipe with attachments running the whole height of the column, the top being the pipe used for a cushion spring. The guides may be bolted to cast-iron brackets embedded in the concrete, or the girt running through the pillars and changeable. With little or no vibration in the battery pillars there would be no fear of it shaking to pieces.

To cushion the vibration of the cam-shaft, the boxing or bearing

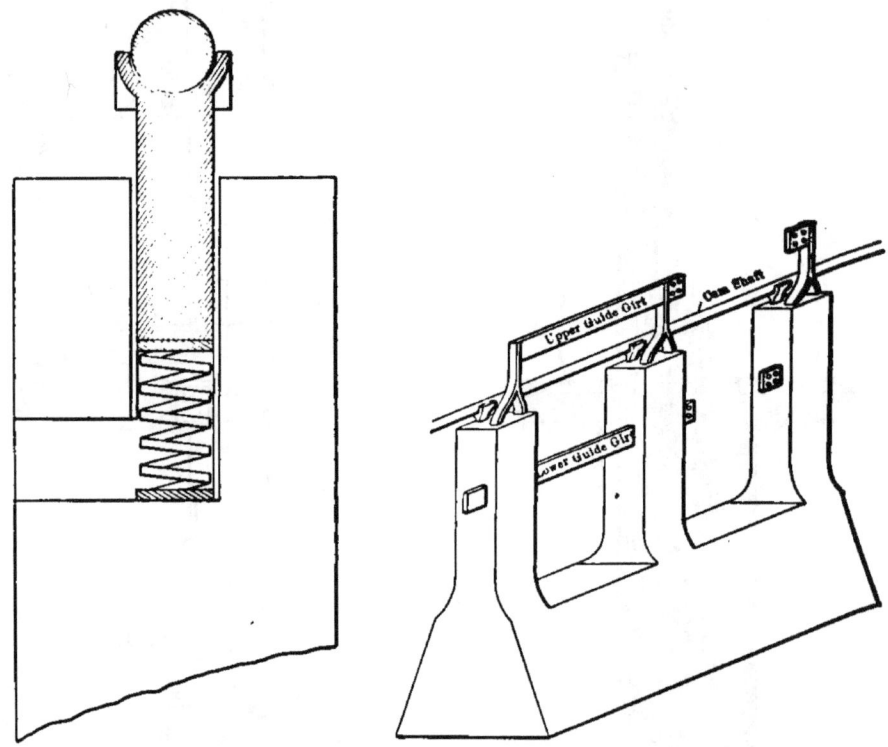

FIG. 50.—Cam-shaft cushion.   FIG. 51.—Concrete battery post.

is a long cast-iron cylinder resting on a heavy railroad carriage spring which in turn is within an 8-in. pipe embedded in the concrete as shown in Fig. 50. An opening is left in the concrete pillar for the purpose of inserting iron shims under the spring to bring the shaft into alignment. The springs must be heavy and with just enough give to take up the vibrations without producing oscillations. Strips of rubber might answer the purpose. Fig. 51 shows the idea of the battery frame.

A description of the concrete structure of the City Deep Mill will show how the concrete battery post has been constructed to withstand the vibrations of a 2000-lb. stamp. Figs. 52 and 53 show the construction in detail, for which I am indebted to Mr. Robeson, the Company's engineer. The mill consists of 200

Fig. 52.—Showing replaceable mortar bolts in an all concrete and steel structure.

stamps each weighing 2000 lb. set back to back. The ore bin is of steel throughout and but little timber is used. Fig. 54, shows the ore bin in process of construction.

The battery posts are of reinforced concrete, 7 ft. high, 4 ft. × 14 in. on the top. On these rest 12 × 14 × 48-in. timbers, and on

7

these is placed the large casting shown in the illustration. The casting supports the cam-shaft along its whole length by means of bearings between each cam. This casting carries the lower guides for the stems, with a wooden cushion between to minimize

FIG. 53.—Leschinger patent cam-shaft supports.

vibrations. The stamp heads are 46 in. long and 9 1/2 in. in diameter, with shoes 14 in. long.

Fig. 55 shows an all steel and concrete construction made by Fraser and Chalmers of England.

Mortars are designed for crushing and for crushing and amalgamating. We may have shallow narrow mortars for fast crushing, or deep narrow mortars to combine crushing with amalgamation, or we may have shallow and deep wide mortars to combine crushing with amalgamation. The character of the ore and the process for the extraction of the precious metals must decide the choice. The narrow mortars do not allow of inside plates, while the wide mortars do, therefore, for in an ore where amalgamation is not followed by any other process it is

Fig. 54.—Concrete and steel structure of the City Deep mill.

important to amalgamate thoroughly, therefore the wide mortar is used that will allow of inside plates. Figs. 56 and 57 show two amalgamating mortars.

Fig. 58 shows a Standard Straight Back Mortar used when it is not desired to amalgamate in the mortar.

Fig. 59 shows a double discharge mortar used for fast crushing.

The standard amalgamating mortar in use in California is 54 in. high, 18 1/2 in. wide at the top and 58 in. long. The width at mortar opening inside of the false lining is 15 1/2 in. and the width at discharge with the screen 6 in. above the die (new) is 17 1/2 in.

This mortar at the Fort Bidwell Mill amalgamated 88 per cent. of the gold in the ore.

The North Star Mortar, Fig. 60, is another form of California

FIG. 55.—Concrete mortar base and all steel frame.

mortar that takes only a chuck-block plate and combines moderately fast crushing with amalgamating.

Fig. 61 shows an open fronted mortar box made by Fraser and Chalmers, Ltd. It has the following advantages:

Easy access to the interior.

Easy replacement and removal of stems, shoes, dies, liners, etc. The removable front can be swung out of and into position

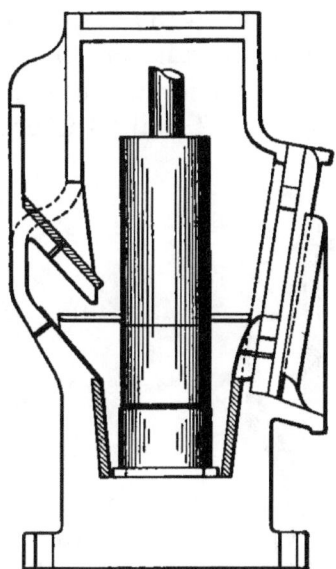

FIG. 56.—Homestake mortar.

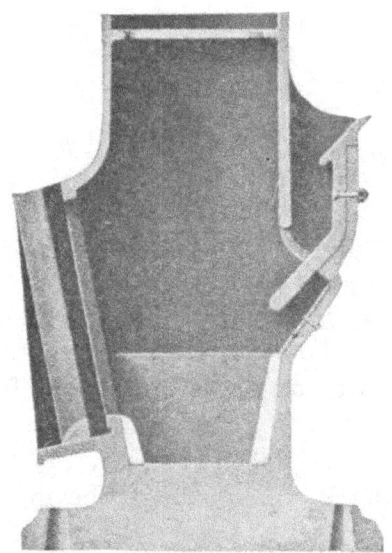

FIG. 57.—Amalgamating mortar.

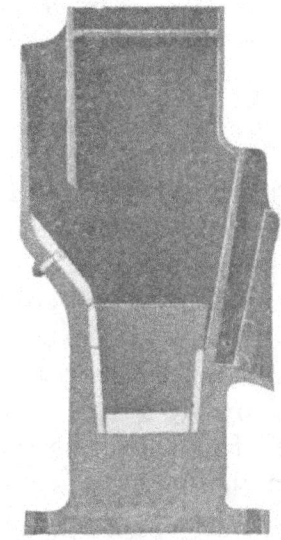

FIG. 58.—Quick crushing mortar.

in a few minutes. Every millman will appreciate the improvement over the usual solid front mortar.

For setting on concrete foundations the base of the mortar is cast to a thickness of 15 in. for the heavy stamps. This does not require an anvil block. Usually the base is 8 to 9 in. thick.

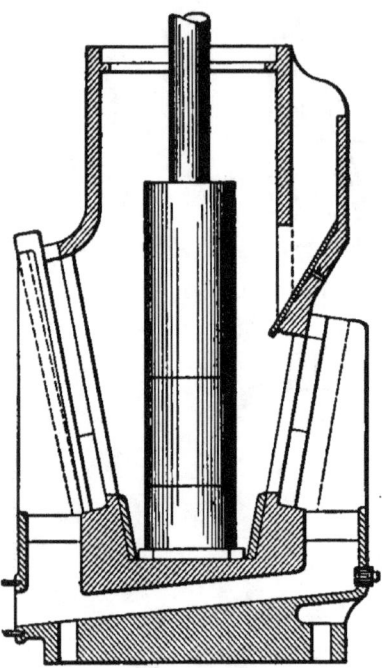

Fig. 59.—Double discharge fast crushing mortar.

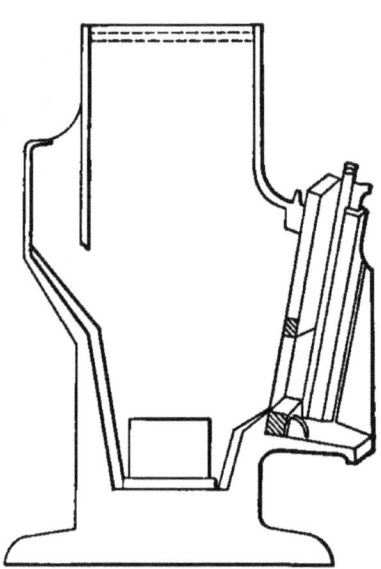

Fig. 60.—North star mortar.

Cams

The proper curve for the face of a cam is the involute of a circle whose radius is the distance from the center of the cam-shaft to the center of the stem; the lift of the cam will then be in the center of the stamp stem, and no pressure thrown on the guides until the toe of the cam has passed the center line of the stem. The distance between the cam-shaft and the stem is therefore fixed by the design of the cam and if this distance is altered by wear of the guides, or the guides are not placed the right distance from the center of cam-shaft the lift will no longer be in the center and friction will result.

The involute form of the cam is supposed to lift the stamp at a uniform speed. The farther the point of contact is from the

Fig. 61.—Open front mortar.

root of the cam the greater the danger from vibrations and therefore the more likelihood of a broken stem.   The radial length of
the cam should not be greater than the lift required; a cam designed for a 7-in. lift should not be used for a 10-in. lift.[1]   Practically it is not possible to design a cam for any size of shaft or stem
that will give all reasonable drops and conform to the conditions
of a perfect load on cam-shaft.

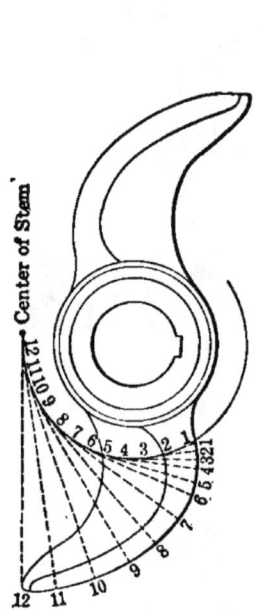

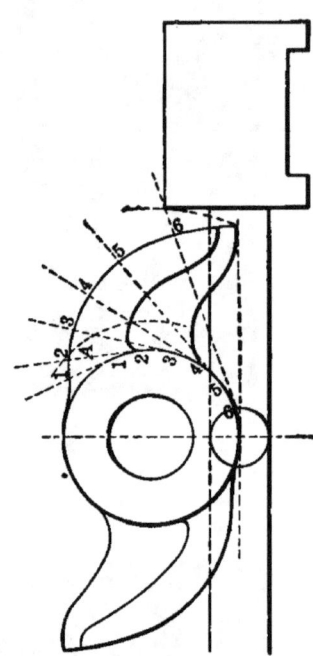

FIG. 62.—Involute cam.　　　　　FIG. 63.—Cam for 8-inch maximum drop.

Fig. 62 shows the method of designing a cam from the involute of a circle.   This is a 16-in. horn cam.

Fig. 63 shows the design for a cam with a 8-in maximum drop.
The horn is 14 in. long and the start of the rise would begin near
the hub of the cam; at the point A.   By practical demonstration
the involute of a circle may be improved upon to correct the
small variations due to variations in the distance between stem
and cam-shaft and variations in the height of drop.

Fig. 64 shows a cam designed by C. O. Smidt for 9 3/ 8 inch drop
which varies somewhat from the involute of a circle and is
said by him to produce a steady rise from start to finish.

Cams are now made of the best cast steel, and when bought

[1] "Gold Mining Machinery," Tinney.

from a responsible dealer will seldom give trouble by breaking. They are now all fastened to the cam-shaft by means of curved gibs which are self tightening and easily detached.

Fig. 65 shows the Canda method of gibbing, Fig. 66 the Blanton, and Fig. 67 the "New" Blanton Patent.

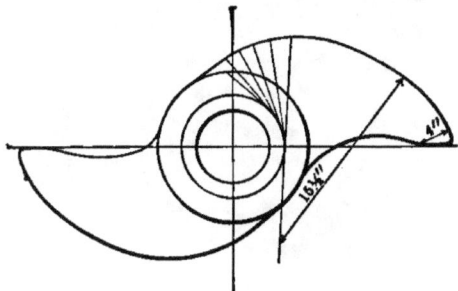

Fig. 64.—Cam for 9 3/8-inch drop. (*Schmidt.*)

In ordering cams the dimensions indicated in the illustration Fig. 68 should be sent to the makers as also the material of which the cam must be made. If the cam-shaft is key-seated and the cams put on with keys and it is desired that the makers cut the key-seat, a template must be sent, but it is usual to cut the key-seats in the mill and order the cams plain. It is also

Fig. 65.—Canda self-fastening cam.

necessary to state whether the cam is a right- or left-hand cam. Fig. 69 shows a ten-stamp battery with right- and left-handed cams. It will be seen that a right-handed cam has its hub on the right when the observer is facing it in such a way that the top is moving from him.

Fig. 66.—Blanton's self-fastening cam.

Fig. 67.—"New" Blanton patent.

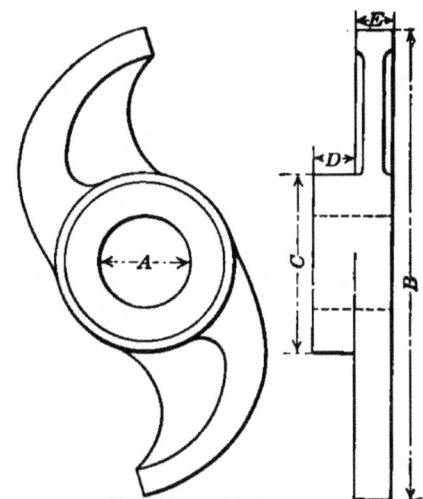

Fig. 68.—Standard cam.

Cam-shafts are designated as right- and left-handed according to which end carries the pulley or bull-wheel. A right-handed cam-shaft may have either right- or left-handed cams just as a left-handed cam-shaft may have left- or right-handed cams.

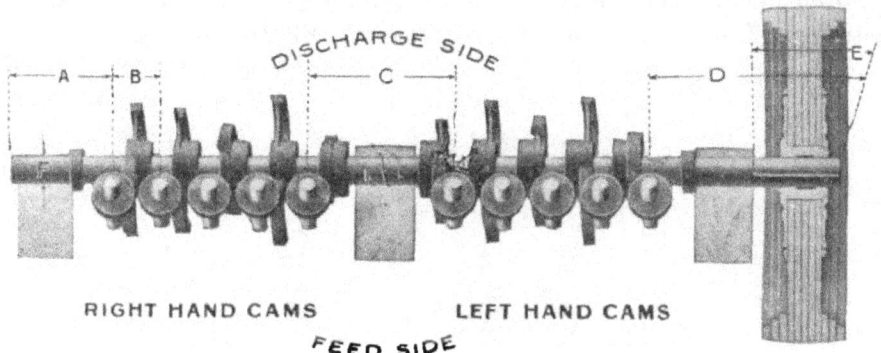

Fig. 69.—Cam-shaft dimensions.

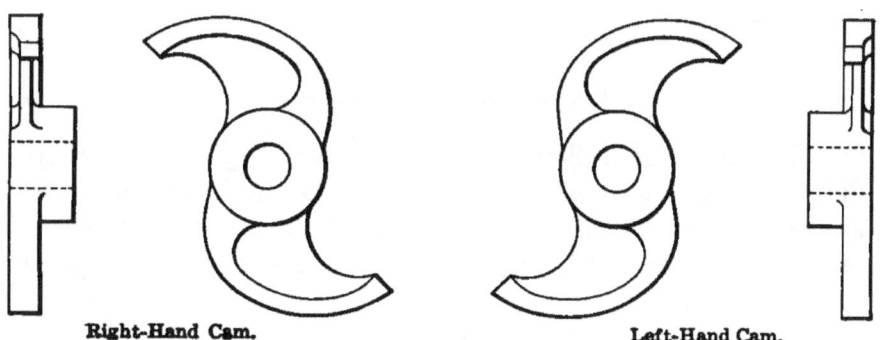

Right-Hand Cam.   Left-Hand Cam.

Fig 70.—Right and left handed cams.

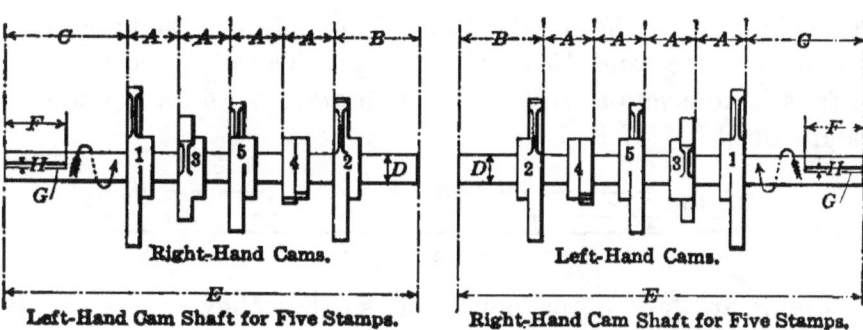

Right-Hand Cams.   Left-Hand Cams.

Left-Hand Cam Shaft for Five Stamps.   Right-Hand Cam Shaft for Five Stamps.

Fig. 71.—Cams and cam-shafts for five stamps.

Fig. 70 shows a right- and left-handed cam, while Fig. 71 shows the dimensions necessary for ordering a five-stamp cam-shaft.

The bull-wheel or driving pulley is always made of **wood** fastened on with a curved gib or wedge as in the self-tightening cams or by means of keys. In ordering a new battery do **not** buy the keyed pulley. The self-tightening ones are so **far** superior in every way and cost nothing extra. They can be taken off easily and never work loose. Those who have worked

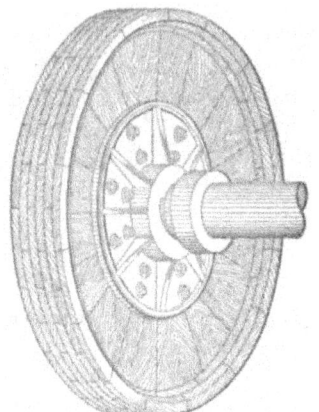

FIG. 72.—Self-tightening bull wheel.

in mills where the pulley is continually working loose know how difficult it is to key one on when the key-seat has become worn and all millmen will appreciate the advantage of a bull-wheel fastened on with a self-tightener. Fig. 72 shows such a pulley.

Stems

Stems are usually made of wrought iron or mild steel, the former to be preferred. The standard size of the California stem is 3 7/16 in. in diameter and 15 to 16 ft. long. Those in use in South Africa in connection with the heavy stamps are 4 in. in diameter and from 13 to 17 ft. long.

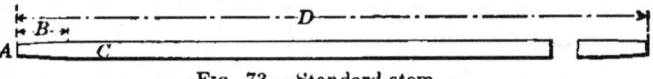

FIG. 73.—Standard stem.

Stems are tapered on both ends so that they are reversible. The taper is usually 1 in. to a foot, or 3/4 in. to a foot, the length of the taper being 5 to 6 in.

Fig. 73 shows the dimensions necessary when ordering new stems, A the diameter at end, B the length of taper, C the diameter of the stem and D the length of stem.

At the North Star Mines Co. mill at Grass Valley, California, the stems are not attached to the boss-heads in the usual manner. They are without taper at either end and are held in the boss by a tapered bushing. This bushing is made of steel and in two pieces, each lacking a sixteenth of an inch of being an entire semi-cylinder, and is grooved to a width of 1/8 in. by a milling saw, from the top almost to the base, in order that the bushing may have a certain amount of spring.

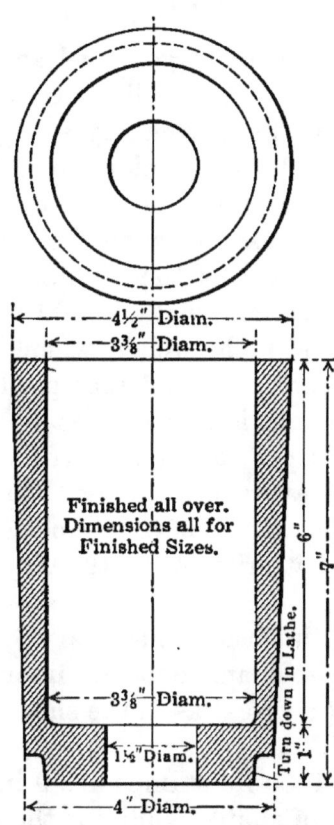

Fig. 74.—Bushing for fastening stem to boss-head.

"The bushing is placed on the stem, then wrapped with a piece of canvas and dropped into the boss with the stem reaching as near to the bottom of the bushing as possible."

When a stem breaks a new boss and shoe are set without taking the stems from the guides. The broken piece of stem is driven out with a drift key, or failing this is shot out with powder. A shoulder is turned in the base of the bushing so any burring of the metal by the drift-key will not project far enough to touch the sides of the boss, and so impede the bushing being driven out.

About 20 per cent. of the bushings are broken in driving the broken pieces of stem out of the boss, but this cost is largely offset by the promptness with which a new boss and shoe can be fitted into a broken stem.   This description and drawing, Fig. 74, is taken from the *Engineering and Mining Journal,* Feb. 25, 1911.

Shoes

Shoes are made of cast iron, forged steel and chrome steel, the last being the best material especially if they come from a distance where freight must be paid in any event.   Where there is a local foundry it may be economical to use cast iron shoes, although even this is doubtful. Chrome steel shoes usually last six months with continual use, although the length of time varies in different mills.   During this time it has crushed about 500 tons of ore.   If the original weight is 200 lb. which is nearly what they weigh new and if the weight when discarded is 40 lb., this gives 160 lb. of wear or about 3 tons per lb. of shoe weight. If the cost including freight is 10 cents per lb. the cost per ton of ore will be 3 1/3 cents per ton of ore crushed.   This is a fair average for an ore of average hardness.   The above estimate refers to a 1000-lb. stamp.   Where a shoe weighing 290 lb. is used as in the heavy batteries in South Africa a different estimate must be made.

Fig. 75 shows the dimensions necessary when ordering new shoes.   The taper of the shank depends upon the taper in the boss-head which is not as great as in the shoe so that the wedges may tighten and hold.

Shoes for the usual 1000-lb. stamps are 9 in. in diameter and 8 to 9 in. high to base of shank, while for the heavy stamps the shoes are 9 1/4 in. in diameter and 14 in. long.

Dies

Dies are made of cast iron, forged steel and chrome steel.   My experience leads to the preference of a steel slightly softer than the chrome steel of the shoe and from one-fourth to one-half greater in diameter.   The former condition gives the wear to the die in preference to the shoe and causes a more even wear of both, preventing cupping. The latter allows for a slight play in the guides, the shoe always hitting on the die. Where both shoe and die are of chrome steel of the same hardness or toughness the wear is liable to be on the shoe more than on the die and an uneven surface is liable to result. For this reason I prefer my shoes of chrome steel and my die of a good grade of forged steel.

Fig. 76 shows the dimensions necessary for ordering dies.

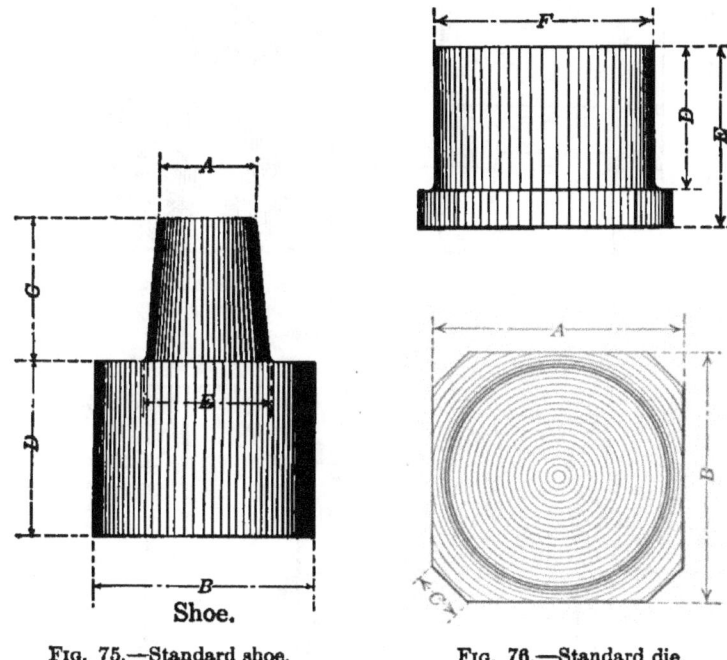

FIG. 75.—Standard shoe.   FIG. 76.—Standard die.

Boss-heads

Boss-heads more commonly called bosses are made of steel and cast iron, the former to be preferred. The steel boss-head is so far superior to the cast iron that the latter should on no account be used. They seldom break while those of cast iron split and chip off on the least occasion.

Fig. 77 shows the dimensions necessary when ordering new bosses. The standard size for a boss is 9 in. in diameter by 18 in. high. The weight varies from 150 to 850 lb. according to the weight of the stamp. In the 1050 lb. stamp the boss weighs about 250 lb. while in the heavy batteries of 1500 to 1600 lb.

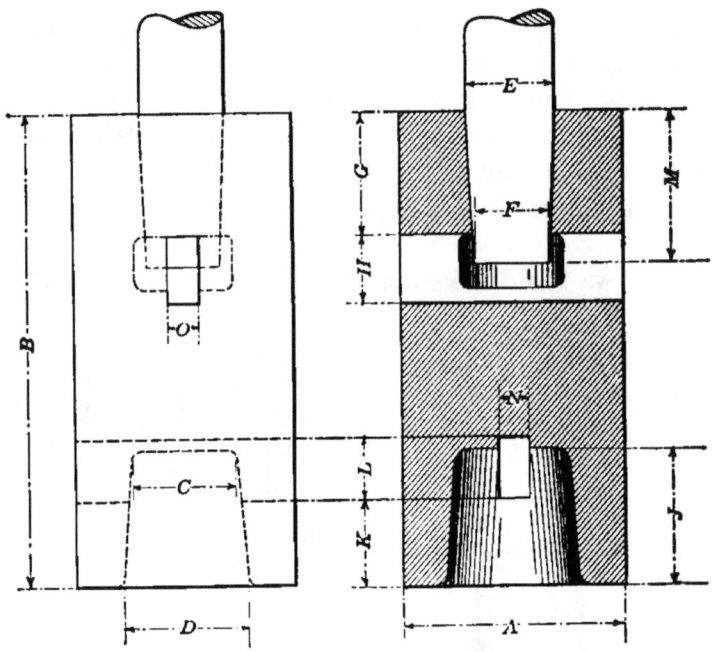

FIG. 77.—Standard boss-head.

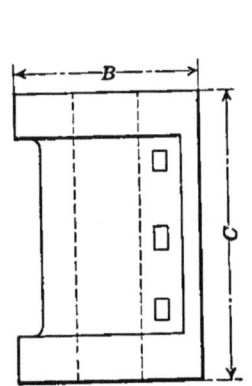

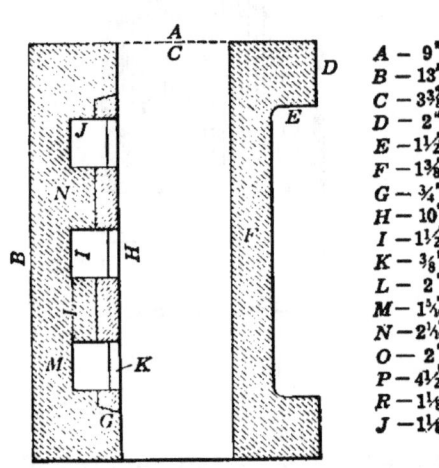

A — 9"
B — 13"
C — 3¾"
D — 2"
E — 1½"
F — 1⅜"
G — ¾"
H — 10"
I — 1½"
K — ⅜"
L — 2"
M — 1¾"
N — 2¼"
O — 2"
P — 4½"
R — 1⅛"
J — 1¼"

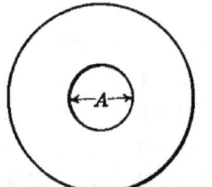

FIG. 78.—Dimensions for three-key tappet.

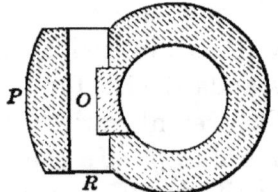

FIG. 79.—Three-key tappet in detail.

they weigh from 350 to 400 lb. In the latter case they are sufficiently long to protrude over the top of the mortar, and are an advantage in this respect that the top is always in sight.

Tappets

Tappets are all made with a gib key, some have two keys, some three and for the extra heavy stamps four keys. Fig. 78 shows

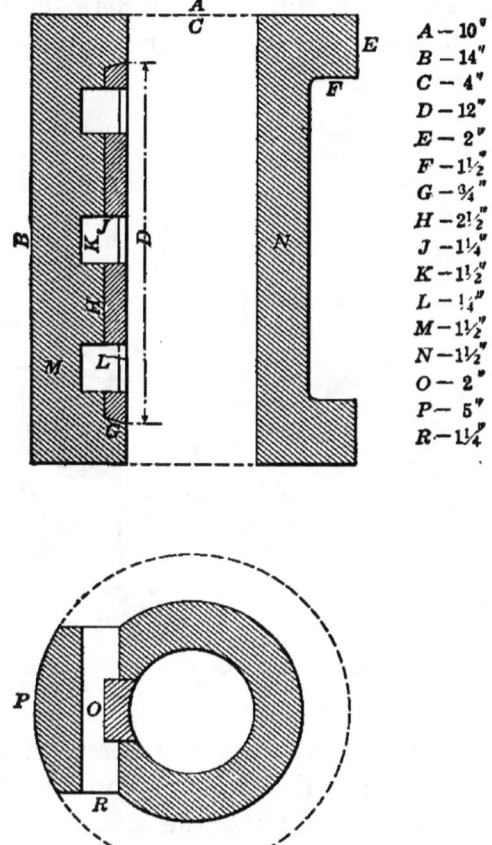

A — 10"
B — 14"
C — 4"
D — 12"
E — 2"
F — 1½"
G — ¾"
H — 2½"
J — 1¼"
K — 1½"
L — 1¼"
M — 1½"
N — 1½"
O — 2"
P — 5"
R — 1¼"

FIG. 80.—Three-key tappet for 4-inch stem.

a two-key tappet with necessary dimensions to furnish the manufacturers if one leaves to their judgment the other dimensions which are important.

Tappets weigh 110 lb. to 280 lb. and are made of steel. Fig. 79 shows the dimensions of a three-key tappet for the standard size stem. Some millmen prefer the key ways convex, I do not because the gib then only bears on the central portion and soon

8

jars loose. The key ways had better be straight. The keys should be made of 1 in. sq. steel tapered but slightly. The more shims used for fitting the keys the more likelihood of them coming loose. The shoulder D should be thick enough to resist pounding with a heavy hammer.

Fig. 80 shows the design of a tappet for a heavy stamp fitted for a 4 in. stem. In all cases the clearance between the stem and the inside of tappet should be as small as possible; a thirty-second of an inch is sufficient. Where the stem has been worn considerable the tappet must be shimmed with sheet iron as described elsewhere.

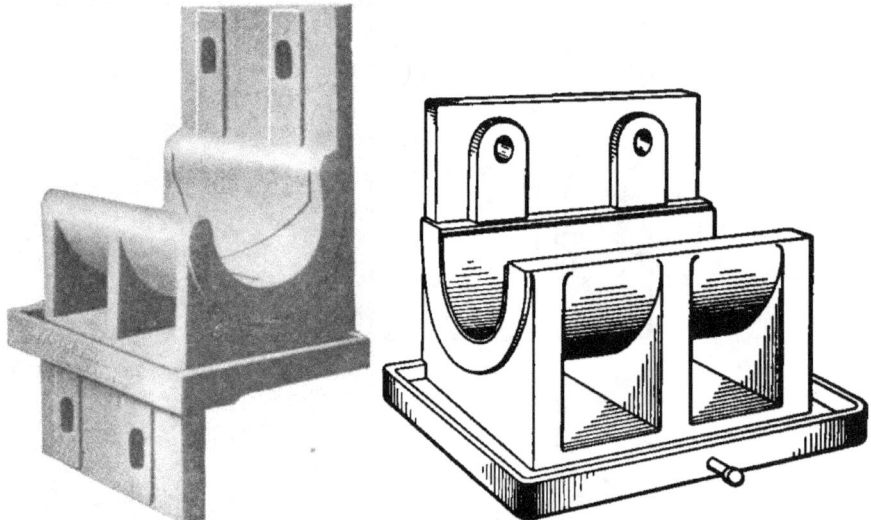

FIG. 82.—Cast-iron cam-shaft box.   FIG. 83.—Standard cam shaft-box. Babbitted or not as required.

Four-key tappets are used on the heavy stamps in South Africa. These weigh 280 lb. or more and are 9 15/16 in. in diameter and 20 in. long, while the usual tappets with the 1050 lb. stamps are 9 in. in diameter and 12 or 13 in. long.

Cam-shaft Boxes

Cam-shaft boxes were formerly all babbited, but of late years the babbit has been discarded and the shaft runs in the cast iron. This is a great improvement and contrary to expectations does not lessen the life of the shaft for if in alignment and greased every three or four hours they cause little trouble either from hot boxes or wear. Figs. 82 and 83 show two shapes of boxes.

To exclude the dust a piece of canvas may be hung over the box. Where one single stamp or two stamps are on one cam-shaft the boxes are sometimes covered with a cap, but this is unnecessary.

Cam-shafts

Cam-shafts are made of chrome nickel steel, Norway iron, re-fined hammered iron and mild steel. The chrome nickel steel is the same material of which automobile axles are made and should be a good material to resist intense vibrations.

It is recorded that cam-shafts of chrome steel were tried at the Boston Con. Mill but broke so frequently that a change was made to malleable iron since and during a period of a year at least, not one broke.

Fig. 84.—The usual cast-iron guide.

Six inches is the usual size of a cam-shaft for the moderate weight stamps and even for lighter batteries they should never be under this diameter.

There is no doubt that a good grade of malleable iron is the best material for a cam-shaft, but machinery houses prefer to furnish a mild steel which so far as my experience goes has given good results.

Stem Guides

Guides are made of wood or cast iron. Sectional wooden guides have never proved much of a success and the long wooden guides are objectionable because to change a stem the whole battery must be held up all the time and the whole guide must be taken off. Where stems do not break frequently the wooden guide made of white oak and 14 in. through the bore, is certainly hard to beat.   Any guide no matter whether of wood or iron must have a long bearing.   The type shown in Fig. 84 has this one objection—that the bearing is only 5 to 6 in. deep.

Figs. 85 and 86 show two long bearing guides which are fre-

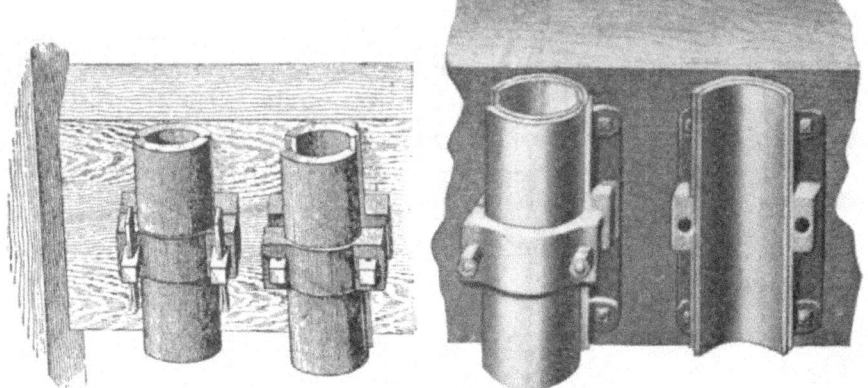

FIG. 85.—Bootless guide.   FIG. 86.—Two-bolt guide.

quently used and give good results.   The use of two bolts one on each side is supposed to allow the stamp to keep in alignment but I think this is more fanciful than real.   Any guide if placed in alignment to start with and not allowed to wear unduly will keep the stamps vertical.

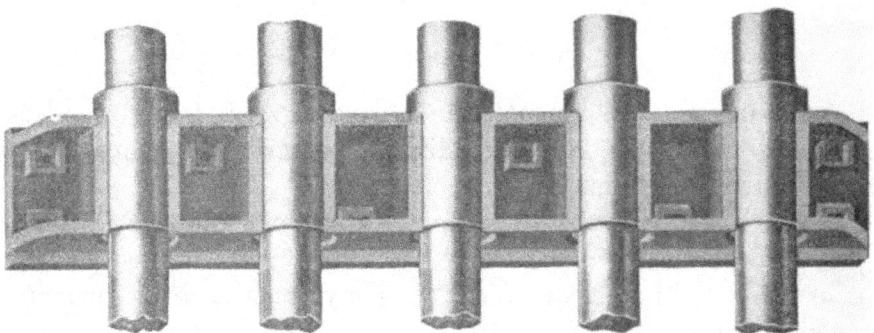

FIG. 87.—Pacific stem guide.

Fig. 87 shows the Pacific guide, probably the best guide made. The only objection is that the shells stick and are sometimes hard to get out, but I am informed that since I had occasion to use them the shape has been changed slightly so that they do not stick.   The shells are removable, are cheap and when worn slightly can be discarded.

The shells for the upper guides are slipped over the stem unless the stem is burred when they are broken in half along a groove in the casting. Those on the lower guides are split before putting on.

Fig. 88 shows the plain long wooden guide, the forerunner of all other guides. At some mills these guides are made of pine, but do not last long and do not pay to use.

The guides cause more trouble as a rule than any other portion of a mill; the bolts break, the nuts come loose, they wear abnormally unless looked after carefully and when worn the least bit the stem wears and the stamp does not fall vertically, thus causing broken stems, perhaps causing a shoe to come off and altogether giving the millman more trouble directly and indirectly than any other part of the mill.

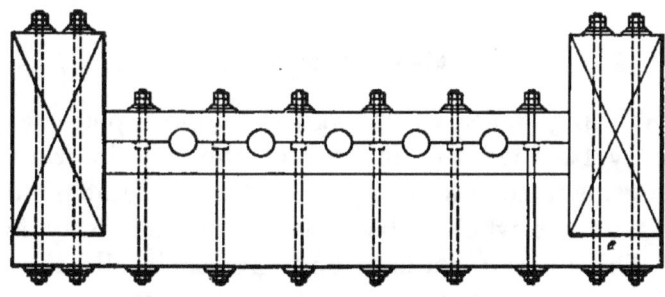

Fig. 88.—Wooden two-piece guide.

The Pacific guide has no nuts or bolts which can loosen and cause trouble. This alone is a great consideration.

To keep guides in good condition they must be greased frequently with compound and not allowed to get dry. The grease must not be put on so lavishly that it falls on the top of the mortar. Joshua Henry Iron Works furnish with their cast iron guides a felt lubricator which not only prevents the grease from going below the guide but being saturated with oil keeps the guide well greased. The guides have a recess horizontally in which these pieces of felt packing fit and cannot come out once the cap of the guide is bolted on.

As many millmen have difficulty in keeping the nuts on the bolts which fasten the cap to the base of the casting the different means employed may be of interest. First there is the double nut with a washer of belting between the nut iron washer. Next some use the two nuts as above described but between the two nuts is a washer of iron or steel cut open so that to screw the second nut tight this washer acts as a spring. This method sometimes succeeds and sometimes not according to the amount of vibration in the battery. Yet another method is with a lock nut, made of two pieces the under one wedging when the upper one is screwed on. I have found that too often they wedge too tight to loosen easily and the threads get worn so that they will not act at all. Another method is to make nuts of a rectangular shape with the hole for thread not in the middle of the nut but to one side so far that the nut may be hammered into place. This long side of the nut being heaviest hangs down and no amount of vibration will cause it to turn around on the bolt. This idea is due I believe to Mr. Courtney De Kalb, at least it was at the exposed Treasure Mill, where he was general manager, that I first used it.

Amalgamating Tables

Amalgamating tables which hold the copper plates are made permanently fastened to the floor of the mill in front of the mortar, are made so that an upper section is detachable, or, are made so that the whole table may be pushed away from the mortar. This latter form is to be preferred in all cases, except where the amalgamating is carried on in a separate section of the mill, when they are of course stationary. For a small mill this system of having the amalgamating department separate from the crushing is out of the question, but for mills of large capacity there is much to be said in its favor. The usual objection is that the pulp does not distribute so well on the plates as from the lip of the mortar. This is easily overcome and in those mills where tried the saving on the plates is as good or better than where the plates are immediately in front of the mortar. There is no doubt that a millman who has to look after as many as 50 stamps and also do the amalgamating must do one or the other an injustice. If his batteries are out of order he must neglect the amalgamating and *vice versa*, also his hands are more or less greasy from handling the machinery and it is almost impossible to have them clean when working on the plates.

There are three important essentials in an amalgamating table, leaving out of consideration the angle of inclination. First, it must be strong and rigid, to bear heavy weights and strains; second, it must be water-tight to prevent loss of gold by leaking, and to prevent loss of temper on the part of the millman who neither desires to see gold lost nor to wade about in rubbers; and, third, it must be level on a line normal to the pitch of the table, or the pulp will go to one side. The center may be a shade lower than the sides, for the tendency is always for the pulp to go to the sides.

Another important consideration is that the floor on which the table rests, or at least, the supports on which the table rests, should be totally detached from the battery itself, for the vibrations caused by the falling stamps, if communicated to the table, will cause amalgam and "quick" to gradually work down

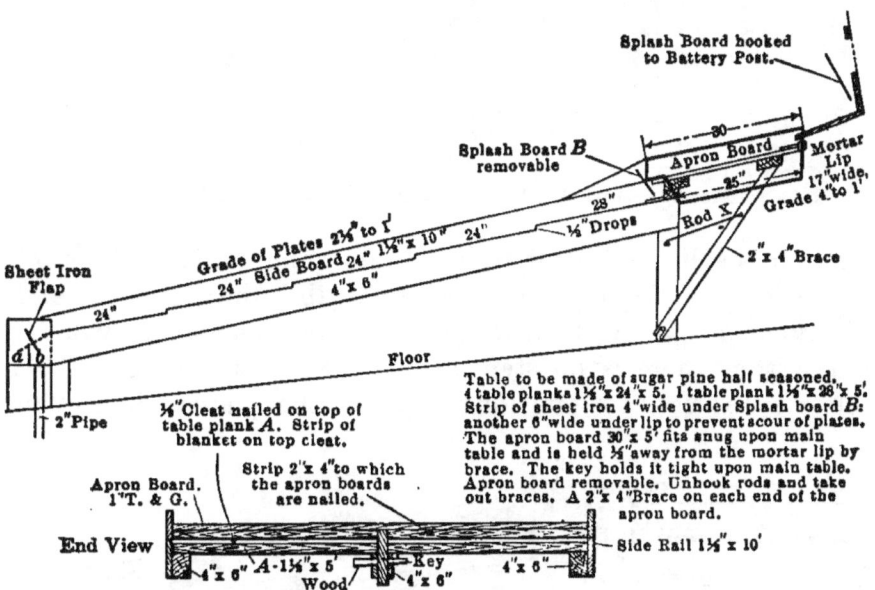

FIG. 89.—Battery palte table, with movable upper section.

the plates. It is also most unpleasant for the millman who has to stand and work on a shaking floor for hours at a stretch. With the old light stamps the vibration was not important, but with heavier stamps such as are now used, it is an import. ant matter and must be reckoned as a primary factor in mill construction.

It is no use describing tables that are stationary in front of the mortar, for no one thinks of erecting one nowadays. Fig. 89 illustrates a partly stationary table, with upper section removable. The trouble is to form a water-tight joint between the removable portion and the stationary part. The table shows five lengths of 2 ft. plates with a step down of 1/2 in. between each plate. The illustration is taken from Dana Harmon and is a good table of this type.

Fig. 90 is the table of the Goldfield Con. Mill and is a splendid example of the removable table. It will be noticed that the con-

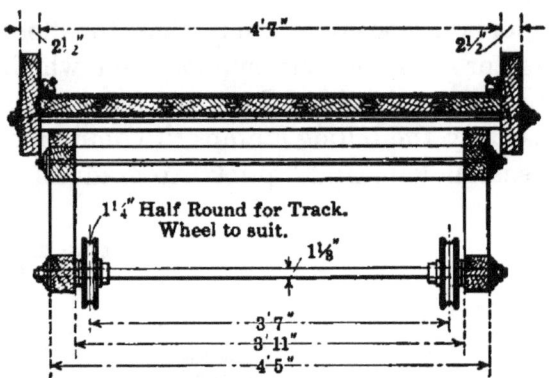

FIG. 90.—Movable table of Combination mill. (*Nevada.*)

struction is solid and a bit elaborate for some mills. The next illustration, Fig. 91, shows a removable table that is easily and quickly made and will answer every requirement. It will be noticed that there is no step down between the plates. This table may be used for one long plate, or the plate if in section

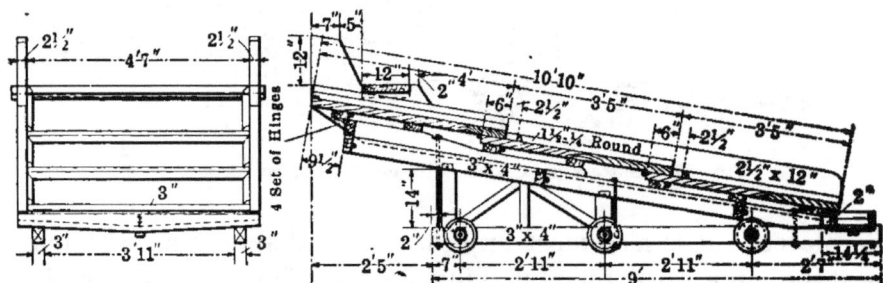

FIG. 90a.—Sectional elevation of "Combination" table.

need only overlap. This will give a fall between the plates of the thickness of the copper.

The side strips, to be sure of a water-tight joint may be bolted down with 1/4 in. bolts, or securely nailed or screwed, as is more often the case. The plates should be underlaid with blanketing or convas.

It is often a matter for discussion whether the top of the table should be made with tight-fitting boards or open spaces left between them. The writer is of the opinion that it makes little difference, if the side strip and all the joints are absolutely water-tight. But can one be sure before actually turning on the water? If there are leakages and the boards are planed to fit tight, any moisture getting under the plates will tend to warp the table and so buckle the plates. A good plan is to have grooved boards

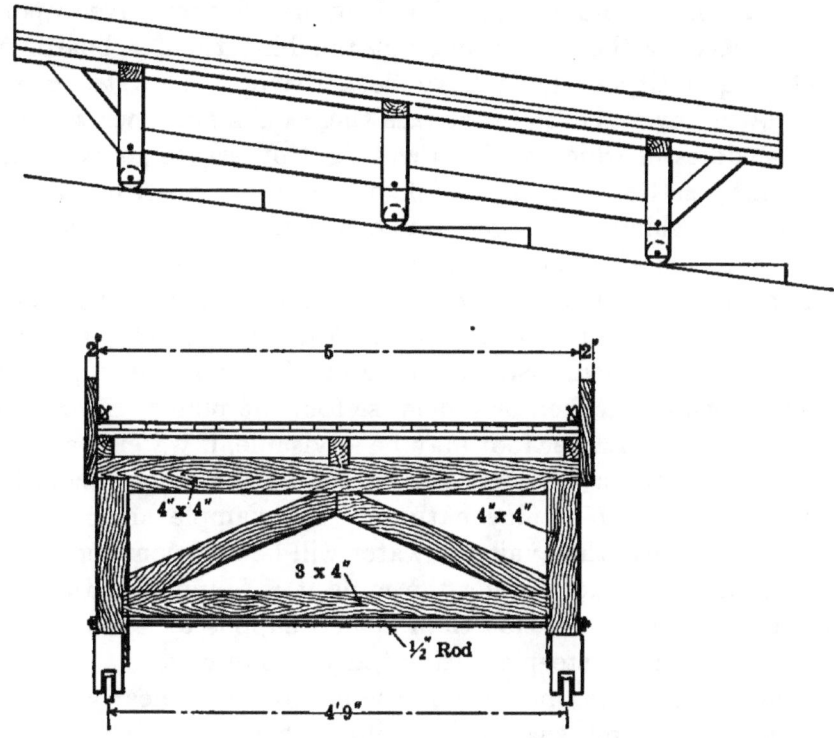

Fig. 91.—Battery plate table.

with a strip inserted into adjoining boards. This is water-tight, but should water leak under the coppers the boards have a chance to expand without buckling. This is shown in the Goldfied Con. Table, Fig. 90. Another method is to leave a space between the boards of about 1/8 in. in which oakum is packed. If water gets under the plates the expansion of the boards will simply press the oakum tighter.

At the end of the table is usually a tail box, the length of which equals the width of the table, and about 8 in. wide.

All the pulp from the table falls into the tail box and thence into the mercury trap.

It is an excellent plan to have riffles on the table. The Yellow Aster Tables have three riffles in the center of the plate made of inch boards shaped Hungarian fashion. The Fort Bidwell Con. Tables have three riffles at the end of the coppers just above the tail box. These riffles catch mercury and amalgam which would otherwise go to the trap where there is a chance of not being caught.

An old scheme and one that Mr. Dimmick has improved upon is to place over the plates and not touching them a board in which a quantity of nails have been driven. The stream of pulp flowing over the plates strikes these nails and goes zig zag leaving eddies in which gold which might otherwise go over the plates may settle. Mr. Dimmick makes these obstructions of rubber and which may then touch the plate without injury.

The grade of the plates depends upon many considerations, such as the constituents of the ore, the amount of water obtainable for milling operations and the thickness of the pulp, and the size of the mesh used. Should there be objectionable sulphides in the ore which blacken or otherwise foul the plates an inclination of 2 to 2 1/2 in. per foot may be advisable, if water is scarce it may be well to have steeper plates than otherwise, particularly where the ore is of a talcy nature. For example an engineer may estimate that the available water will be sufficient for a ten stamp mill but on developing the mine it is found that the ore is more of the nature of clay than at first supposed. The plates could then be made steeper without any loss in gold.

Should the pulp be thick, say in the proportion of nearly one of water to one of sand, such as is discharged from a tube mill the plates must be sufficiently steep to take this sand off. A grade of 2 1/2 in. per foot is sufficient for this purpose.

The mesh of the battery screen will also determine to a certain extent the grade of the plates for large particles of sand either require a steeper grade or more water. I believe as a rule plates could be used steeper than is the usual custom without any loss in extraction. The grade should never be under 1 3/4 in. per foot and may be as high as 2 3/4 in. Some millmen will favor plates never being under 2 in. per foot, but I believe there are cases where a lower grade than this will be found correct for that ore. It is a matter of experiment and those that never try anything new except what others in the district have determined upon will never know the best conditions for catching their gold.

WEIGHTS AND SIZES OF VARIOUS PARTS FOR VARIOUS WEIGHTS OF STAMPS

	750	850	1050	1250	1550	1650	2000
Weight of stamps, pounds	750	850	1050	1250	1550	1650	2000
Weight of shoes	115	140	145–168	160–175	286	286	290
Weight of boss head	182	182	270–365	360–365	376	408	872
Weight of stem	352	435	400–465	575–580	726	726	556
Weight of tappet	110	110	120–150	135–145	162	250	282
Size of shoe		9x8	9x9	9 1/4x14	9 1/4x14		9 1/4x14
Size of boss		9x18	9x18	9 1/4x24	9 1/2x24		9 1/2x50
Size of stem	3 7/16x16		3 1/2x18	4x17	4x17		4x13
Size of tappet		9x13	9 1/4x13	9 5/8x13	10 1/8x18		9 15/16x20
Space between stems	10	10	10	10 1/4	10 1/4	10 3/4	10 3/4
Diameter cam shaft	5	5 1/2	6	6	6 1/2	7	7
Distance between bearings, inches	56	57	58	77	77	78	77
Thickness of base of mortar, inches	6	7	8 1/2	9	9	15	11

The three last columns are taken from C. O. Smidt and represent mills in South Africa.

The usual California mill of 1050 lb. shows quite a range in the weights of the parts. One maker specifies the weights as follows:

Chrome steel boss head	275
Chrome steel tappet	147
Chrome steel shoe	168
Mild steel stem	465
	1050

Ore Feeders

Ore feeders regulate the amount of ore going to the battery and therefore influence to a certain extent the capacity of the mill. A feeder that is intermittent in its action sometimes letting ore down and sometimes not under the same pressure on the lever will not feed the stamps regularly and so at times the stamps will have too much ore and sometimes too little ore for good work.

The Tullock feeder which was in use years ago has now gone out of date and the most usual type seen is the Challenge, of which there are many variations, but all have the revolving plate. In the usual Challenge feeder the circular motion of the revolving plate is caused by the stamp hitting the feeder arm which in turn rotates the friction-wheel. The plate is prevented from returning by the pawls in the friction-wheel. The improvements which are of importance are suspending the feeders so that the space under it is clear and can be readily cleaned with a shovel

Fig. 92.

and replacing the pawls with a friction clutch to get rid of the inconvenience due to pawls dropping out.

Fig. 92 shows such an improvement made by the Hendy Iron Works of San Francisco. The hopper plate is revolved by means of a steel cable on the end of which is a spring. The action is easily seen. The bumper or feeder rod when hit by the falling stamp pulls this rope against the spring which when the stamp has risen pulls the rope back leaving the plate stationary.

Fig. 93 shows the Colorado Iron Works' "Perfect" Challenge feeder. All gears are dispensed with, the motion of the bumper rod being transmitted to the clutch through flexible connected levers.

The ordinary pawl revolving Challenge feeder may be improved

Fig. 93.—Colorado Iron Works' feeder.

Fig. 94.—Prospecting battery.

upon by discarding the pawls and putting in their place a friction grip acting on the outside of the friction-wheel.

Prospecting Mills

Prospecting mills are usually of the single unit pattern, although sometimes five stamps of light weight are in use. These latter are much to be preferred because the prospector usually has an oxidized, free milling ore to treat and it is important that he catch the maximum amount on his plates. As before stated the single unit batteries are crushing batteries and do not give good

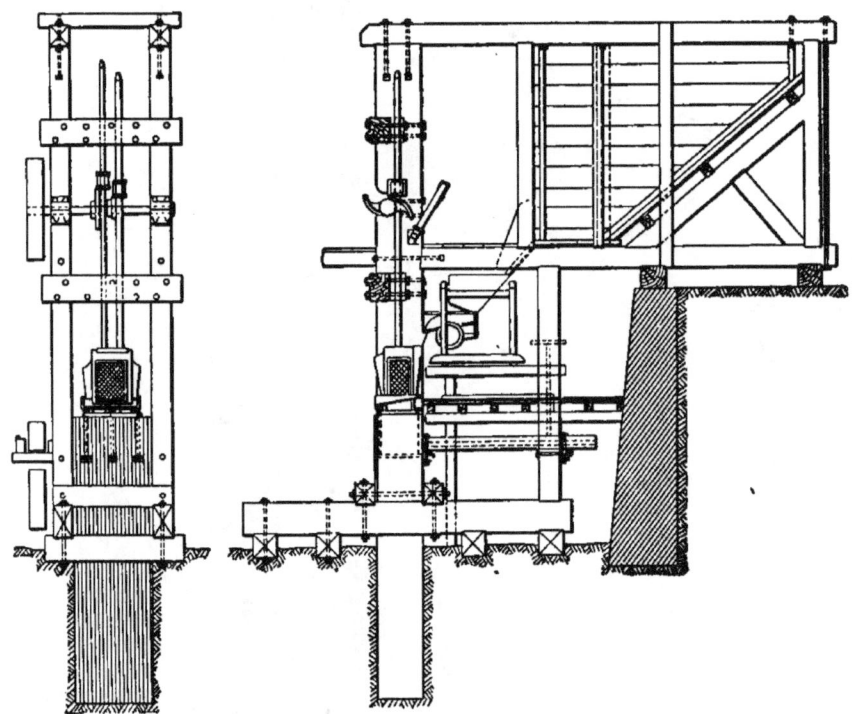

FIG. 95.—Two-stamps standard mill.

results amalgamating so that for a prospector these single units are not satisfactory. Should he not be able to afford a small five-stamp mill he must do the next best and be satisfied with a single stamp.

Fig. 94 is a Fraser and Chalmer Ltd. three-stamp outfit which would make a good mill for a man of small means.

Fig. 95 is a standard weight two-stamp battery. They are poor amalgamators but may be replaced at any time by a standard

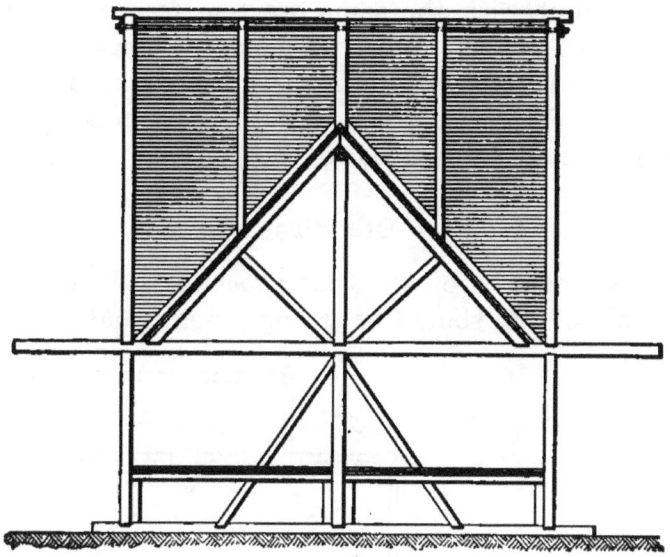

Fig. 96.—Back to back ore bin.

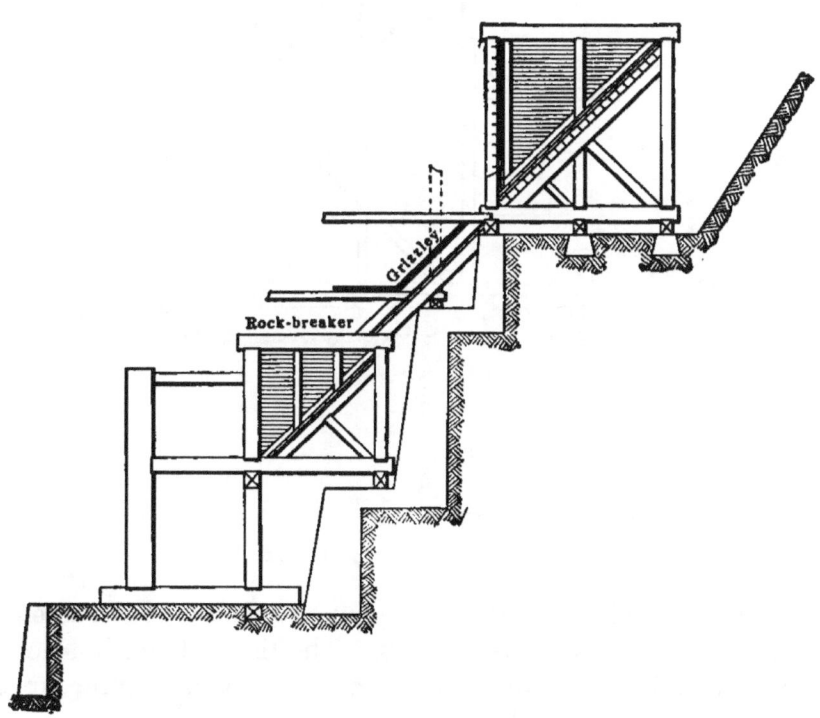

Fig. 97.—Two-bin construction.

five stamp unit with the same ore bin, feeder, battery posts, stamps, cams, tappets, etc. The mortar block is a little too small both in length and width for a five-stamp unit, but when erecting the mill this can be provided for by making the dimensions large enough for the change.

Ore Bins

Ore bins should be of sufficient capacity to hold at least two days supply of ore, should be strongly made and well anchored.

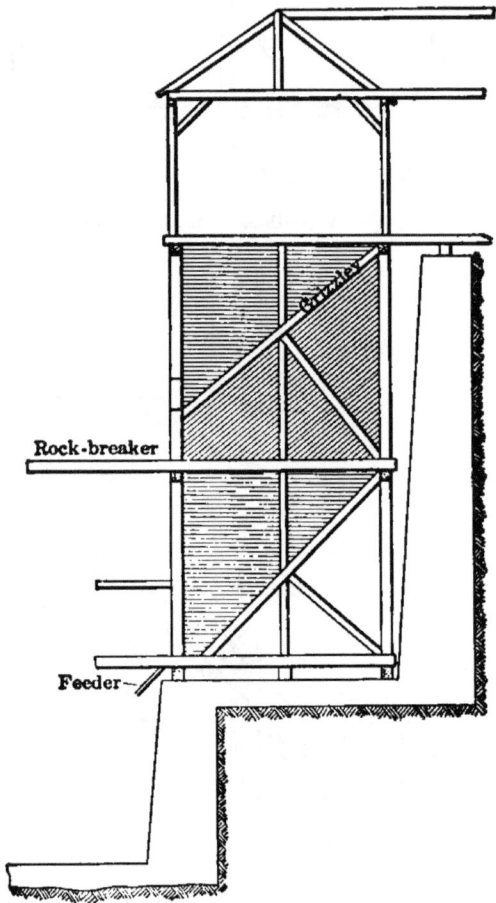

FIG. 98.—Southern cross bin.

Both inclined and flat-bottom bins are now constructed according to individual preferences. The flat-bottom bin holds a reserve supply of ore, but has the disadvantage of tying up this amount of ore which when needed must be shovelled at an extra expense. The best plan is an inclined bottom with just sufficient inclination to hold enough ore so that when the ore has been drawn out by gravity there will be just enough left to protect the bottom from the wear incident to dumping cars of ore from the top of the bin.

If the battery is of the back knee pattern the sill of the ore bin projects to the battery posts, where the mortice in the battery post should be 2 in. larger in the vertical dimension than the tongue on the ore bin timber. Should the battery post or the ore bin settle the sill will have 2 in. play in a vertical direction and will not have to hold the weight of the battery post. This sill or the back knee is for the purpose of holding in place and steadying the battery post and is bolted to the battery post by a rod from the post to the ore bin timbers. A good plan is to erect the bin independent of the battery post forming the back knee with a 12 × 12 timber which sets in a cast-iron shoe or bracket on both battery post and ore bin. Should the battery post now get out of the vertical this timber may be planed off or shimmed according to requirements. In the front knee construction, the ore bin is independent of the battery.

Fig. 96 shows an ore bin for the back to back type of mill. This is not usual in small mills, but may be found in mills of 80 stamps and upward.

Fig. 97 shows an excellent arrangement for mills of any size. The upper bin, the ore storage bin. The grizzley may be at the top of this bin or as shown at the lower end. With this arrangement it is not necessary for the rockbreaker man to be always at the breaker. He may allow the ore to accumulate while he is engaged in other work.

Fig. 98 is a different idea with a storage bin for the coarse ore not going through the grizzley, and is suitable for any size mill. The fine ore storage bin is of great capacity, an excellent idea.

9

INDEX

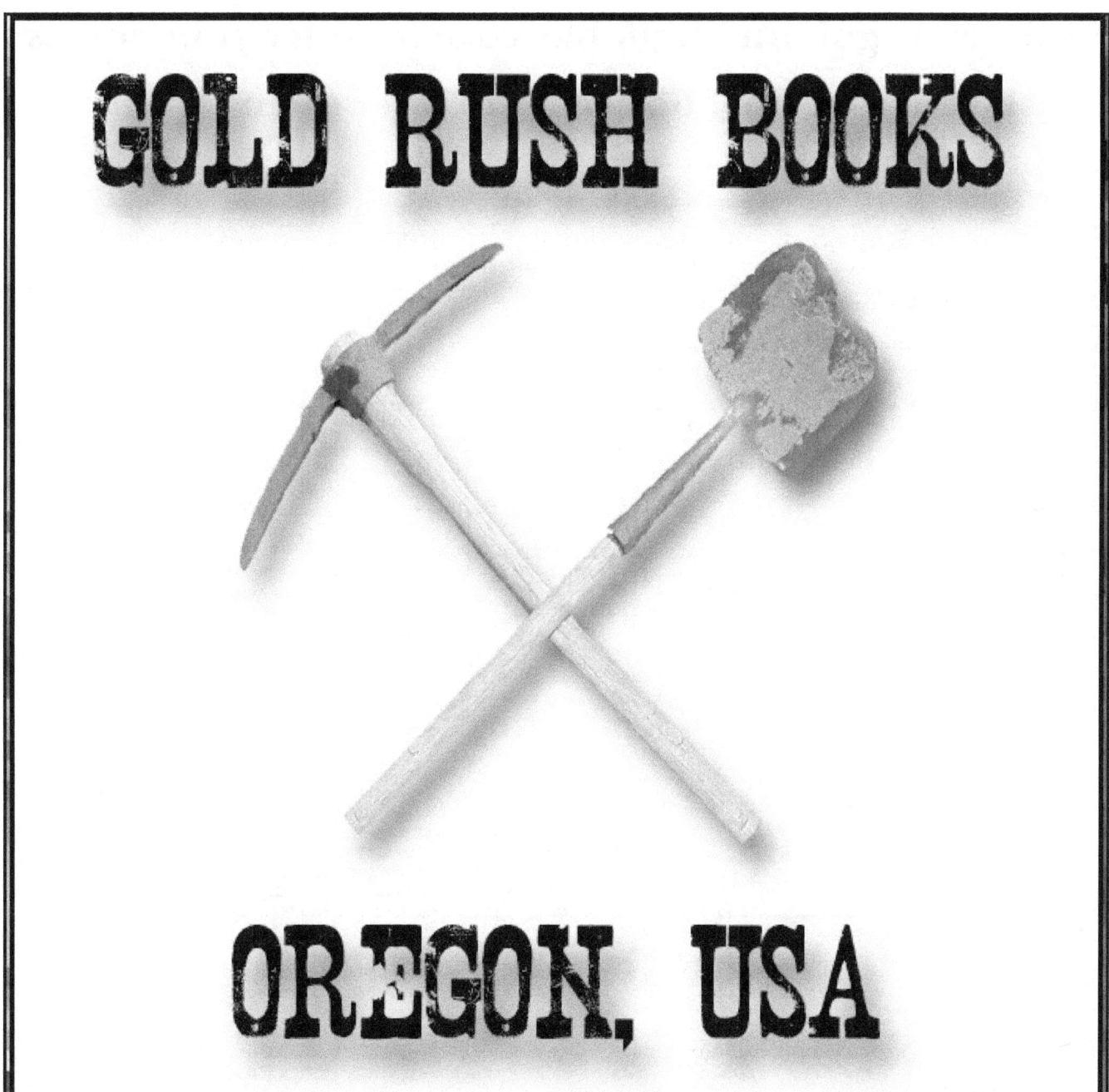

GOLD RUSH BOOKS

OREGON, USA

www.GoldMiningBooks.com

Books On Mining

Visit: www.goldminingbooks.com to order your copies or ask your favorite book seller to offer them.

Mining Books by Kerby Jackson

<u>Gold Dust: Stories From Oregon's Mining Years</u> - Oregon mining historian and prospector, Kerby Jackson, brings you a treasure trove of seventeen stories on Southern Oregon's rich history of gold prospecting, the prospectors and their discoveries, and the breathtaking areas they settled in and made homes. 5" X 8", 98 ppgs. Retail Price: $11.99

<u>The Golden Trail: More Stories From Oregon's Mining Years</u> - In his follow-up to "Gold Dust: Stories of Oregon's Mining Years", this time around, Jackson brings us twelve tales from Oregon's Gold Rush, including the story about the first gold strike on Canyon Creek in Grant County, about the old timers who found gold by the pail full at the Victor Mine near Galice, how Iradel Bray discovered a rich ledge of gold on the Coquille River during the height of the Rogue River War, a tale of two elderly miners on the hunt for a lost mine in the Cascade Mountains, details about the discovery of the famous Armstrong Nugget and others. 5" X 8", 70 ppgs. Retail Price: $10.99

Oregon Mining Books

<u>Geology and Mineral Resources of Josephine County, Oregon</u> - Unavailable since the 1970's, this important publication was originally compiled by the Oregon Department of Geology and Mineral Industries and includes important details on the economic geology and mineral resources of this important mining area in South Western Oregon. Included are notes on the history, geology and development of important mines, as well as insights into the mining of gold, copper, nickel, limestone, chromium and other minerals found in large quantities in Josephine County, Oregon. 8.5" X 11", 54 ppgs. Retail Price: $9.99

<u>Mines and Prospects of the Mount Reuben Mining District</u> - Unavailable since 1947, this important publication was originally compiled by geologist Elton Youngberg of the Oregon Department of Geology and Mineral Industries and includes detailed descriptions, histories and the geology of the Mount Reuben Mining District in Josephine County, Oregon. Included are notes on the history, geology, development and assay statistics, as well as underground maps of all the major mines and prospects in the vicinity of this much neglected mining district. 8.5" X 11", 48 ppgs. Retail Price: $9.99

<u>The Granite Mining District</u> - Notes on the history, geology and development of important mines in the well known Granite Mining District which is located in Grant County, Oregon. Some of the mines discussed include the Ajax, Blue Ribbon, Buffalo, Continental, Cougar-Independence, Magnolia, New York, Standard and the Tillicum. Also included are many rare maps pertaining to the mines in the area. 8.5" X 11", 48 ppgs. Retail Price: $9.99

<u>Ore Deposits of the Takilma and Waldo Mining Districts of Josephine County, Oregon</u> - The Waldo and Takilma mining districts are most notable for the fact that the earliest large scale mining of placer gold and copper in Oregon took place in these two areas. Included are details about some of the earliest large gold mines in the state such as the Llano de Oro, High Gravel, Cameron, Platerica, Deep Gravel and others, as well as copper mines such as the famous Queen of Bronze mine, the Waldo, Lily and Cowboy mines. This volume also includes six maps and 20 original illustrations. 8.5" X 11", 74 ppgs. Retail Price: $9.99

<u>Metal Mines of Douglas, Coos and Curry Counties, Oregon</u> - Oregon mining historian Kerby Jackson introduces us to a classic work on Oregon's mining history in this important re-issue of Bulletin 14C Volume 1, otherwise known as the Douglas, Coos & Curry Counties, Oregon Metal Mines Handbook. Unavailable since 1940, this important publication was originally compiled by the Oregon Department of Geology and Mineral Industries includes detailed descriptions, histories and the geology of over 250 metallic mineral mines and prospects in this rugged area of South West Oregon. 8.5" X 11", 158 ppgs. Retail Price: $19.99

Metal Mines of Jackson County, Oregon - Unavailable since 1943, this important publication was originally compiled by the Oregon Department of Geology and Mineral Industries includes detailed descriptions, histories and the geology of over 450 metallic mineral mines and prospects in Jackson County, Oregon. Included are such famous gold mining areas as Gold Hill, Jacksonville, Sterling and the Upper Applegate. **8.5" X 11", 220 ppgs. Retail Price: $24.99**

Metal Mines of Josephine County, Oregon - Oregon mining historian Kerby Jackson introduces us to a classic work on Oregon's mining history in this important re-issue of Bulletin 14C, otherwise known as the Josephine County, Oregon Metal Mines Handbook. Unavailable since 1952, this important publication was originally compiled by the Oregon Department of Geology and Mineral Industries includes detailed descriptions, histories and the geology of over 500 metallic mineral mines and prospects in Josephine County, Oregon. **8.5" X 11", 250 ppgs. Retail Price: $24.99**

Metal Mines of North East Oregon - Oregon mining historian Kerby Jackson introduces us to a classic work on Oregon's mining history in this important re-issue of Bulletin 14A and 14B, otherwise known as the North East Oregon Metal Mines Handbook. Unavailable since 1941, this important publication was originally compiled by the Oregon Department of Geology and Mineral Industries and includes detailed descriptions, histories and the geology of over 750 metallic mineral mines and prospects in North Eastern Oregon. **8.5" X 11", 310 ppgs. Retail Price: $29.99**

Metal Mines of North West Oregon - Oregon mining historian Kerby Jackson introduces us to a classic work on Oregon's mining history in this important re-issue of Bulletin 14D, otherwise known as the North West Oregon Metal Mines Handbook. Unavailable since 1951, this important publication was originally compiled by the Oregon Department of Geology and Mineral Industries and includes detailed descriptions, histories and the geology of over 250 metallic mineral mines and prospects in North Western Oregon. **8.5" X 11", 182 ppgs. Retail Price: $19.99**

Mines and Prospects of Oregon - Mining historian Kerby Jackson introduces us to a classic mining work by the Oregon Bureau of Mines in this important re-issue of The Handbook of Mines and Prospects of Oregon. Unavailable since 1916, this publication includes important insights into hundreds of gold, silver, copper, coal, limestone and other mines that operated in the State of Oregon around the turn of the 19th Century. Included are not only geological details on early mines throughout Oregon, but also insights into their history, production, locations and in some cases, also included are rare maps of their underground workings. **8.5" X 11", 314 ppgs. Retail Price: $24.99**

Lode Gold of the Klamath Mountains of Northern California and South West Oregon
(See California Mining Books)

Mineral Resources of South West Oregon - Unavailable since 1914, this publication includes important insights into dozens of mines that once operated in South West Oregon, including the famous gold fields of Josephine and Jackson Counties, as well as the Coal Mines of Coos County. Included are not only geological details on early mines throughout South West Oregon, but also insights into their history, production and locations. **8.5" X 11", 154 ppgs. Retail Price: $11.99**

Chromite Mining in The Klamath Mountains of California and Oregon
(See California Mining Books)

Southern Oregon Mineral Wealth - Unavailable since 1904, this rare publication provides a unique snapshot into the mines that were operating in the area at the time. Included are not only geological details on early mines throughout South West Oregon, but also insights into their history, production and locations. Some of the mining areas include Grave Creek, Greenback, Wolf Creek, Jump Off Joe Creek, Granite Hill, Galice, Mount Reuben, Gold Hill, Galls Creek, Kane Creek, Sardine Creek, Birdseye Creek, Evans Creek, Foots Creek, Jacksonville, Ashland, the Applegate River, Waldo, Kerby and the Illinois River, Althouse and Sucker Creek, as well as insights into local copper mining and other topics. **8.5" X 11", 64 ppgs. Retail Price: $8.99**

Geology and Ore Deposits of the Takilma and Waldo Mining Districts - Unavailable since the 1933, this publication was originally compiled by the United States Geological Survey and includes details on gold and copper mining in the Takilma and Waldo Districts of Josephine County, Oregon. The Waldo and Takilma mining districts are most notable for the fact that the earliest large scale mining of placer gold and copper in Oregon took place in these two areas. Included in this report are details about some of the earliest large gold mines in the state such as the Llano de Oro, High Gravel, Cameron, Platerica, Deep Gravel and others, as well as copper mines such as the famous Queen of Bronze mine, the Waldo, Lily and Cowboy mines. In addition to geological examinations, insights are also provided into the production, day to day operations and early histories of these mines, as well as calculations of known mineral reserves in the area. This volume also includes six maps and 20 original illustrations. **8.5" X 11", 74 ppgs. Retail Price: $9.99**

Gold Mines of Oregon - Oregon mining historian Kerby Jackson introduces us to a classic work on Oregon's mining history in this important re-issue of Bulletin 61, otherwise known as "Gold and Silver In Oregon". Unavailable since 1968, this important publication was originally compiled by geologists Howard C. Brooks and Len Ramp of the Oregon Department of Geology and Mineral Industries and includes detailed descriptions, histories and the geology of over 450 gold mines Oregon. Included are notes on the history, geology and gold production statistics of all the major mining areas in Oregon including the Klamath Mountains, the Blue Mountains and the North Cascades. While gold is where you find it, as every miner knows, the path to success is to prospect for gold where it was previously found. **8.5" X 11", 344 ppgs. Retail Price: $24.99**

Mines and Mineral Resources of Curry County Oregon - Originally published in 1916, this important publication on Oregon Mining has not been available for nearly a century. Included are rare insights into the history, production and locations of dozens of gold mines in Curry County, Oregon, as well as detailed information on important Oregon mining districts in that area such as those at Agness, Bald Face Creek, Mule Creek, Boulder Creek, China Diggings, Collier Creek, Elk River, Gold Beach, Rock Creek, Sixes River and elsewhere. Particular attention is especially paid to the famous beach gold deposits of this portion of the Oregon Coast. **8.5" X 11", 140 ppgs. Retail Price: $11.99**

Chromite Mining in South West Oregon - Originally published in 1961, this important publication on Oregon Mining has not been available for nearly a century. Included are rare insights into the history, production and locations of nearly 300 chromite mines in South Western Oregon. **8.5" X 11", 184 ppgs. Retail Price: $14.99**

Mineral Resources of Douglas County Oregon - Originally published in 1972, this important publication on Oregon Mining has not been available for nearly forty years. Included are rare insights into the geology, history, production and locations of numerous gold mines and other mining properties in Douglas County, Oregon. **8.5" X 11", 124 ppgs. Retail Price: $11.99**

Mineral Resources of Coos County Oregon - Originally published in 1972, this important publication on Oregon Mining has not been available for nearly forty years. Included are rare insights into the geology, history, production and locations of numerous gold mines and other mining properties in Coos County, Oregon. **8.5" X 11", 100 ppgs. Retail Price: $11.99**

Mineral Resources of Lane County Oregon - Originally published in 1938, this important publication on Oregon Mining has not been available for nearly seventy five years. Included are extremely rare insights into the geology and mines of Lane County, Oregon, in particular in the Bohemia, Blue River, Oakridge, Black Butte and Winberry Mining Districts. **8.5" X 11", 82 ppgs. Retail Price: $9.99**

Mineral Resources of the Upper Chetco River of Oregon: Including the Kalmiopsis Wilderness - Originally published in 1975, this important publication on Oregon Mining has not been available for nearly forty years. Withdrawn under the 1872 Mining Act since 1984, real insight into the minerals resources and mines of the Upper Chetco River has long been unavailable due to the remoteness of the area. Despite this, the decades of battle between property owners and environmental extremists over the last private mining inholding in the area has continued to pique the interest of those interested in mining and other forms of natural resource use. Gold mining began in the area in the 1850's and has a rich history in this geographic area, even if the facts surrounding it are little known. Included are twenty two rare photographs, as well as insights into the Becca and Morning Mine, the Emmly Mine (also known as Emily Camp), the Frazier Mine, the Golden Dream or Higgins Mine, Hustis Mine, Peck Mine and others. **8.5" X 11", 64 ppgs. Retail Price: $8.99**

Gold Dredging in Oregon - Originally published in 1939, this important publication on Oregon Mining has not been available for nearly seventy five years. Included are extremely rare insights into the history and day to day operations of the dragline and bucketline gold dredges that once worked the placer gold fields of South West and North East Oregon in decades gone by. Also included are details into the areas that were worked by gold dredges in Josephine, Jackson, Baker and Grant counties, as well as the economic factors that impacted this mining method. This volume also offers a unique look into the values of river bottom land in relation to both farming and mining, in how farm lands were mined, re-soiled and reclamated after the dredges worked them. Featured are hard to find maps of the gold dredge fields, as well as rare photographs from a bygone era. **8.5" X 11", 86 ppgs. Retail Price: $8.99**

Quick Silver Mining in Oregon - Originally published in 1963, this important publication on Oregon Mining has not been available for over fifty years. This publication includes details into the history and production of Elemental Mercury or Quicksilver in the State of Oregon. **8.5" X 11", 238 ppgs. Retail Price: $15.99**

Mines of the Greenhorn Mining District of Grant County Oregon - Originally published in 1948, this important publication on Oregon Mining has not been available for over sixty five years. In this publication are rare insights into the mines of the famous Greenhorn Mining District of Grant County, Oregon, especially the famous Morning Mine. Also included are details on the Tempest, Tiger, Bi-Metallic, Windsor, Psyche, Big Johnny, Snow Creek, Banzette and Paramount Mines, as well as prospects in the vicinities in the famous mining areas of Mormon Basin, Vinegar Basin and Desolation Creek. Included are hard to find mine maps and dozens of rare photographs from the bygone era of Grant County's rich mining history. **8.5" X 11", 72 ppgs. Retail Price: $9.99**

Geology of the Wallowa Mountains of Oregon: Part I (Volume 1) - Originally published in 1938, this important publication on Oregon Mining has not been available for nearly seventy five years. Included are details on the geology of this unique portion of North Eastern Oregon. This is the first part of a two book series on the area. Accompanying the text are rare photographs and historic maps.8.5" X 11", 92 ppgs. Retail Price: $9.99

Geology of the Wallowa Mountains of Oregon: Part II (Volume 2) - Originally published in 1938, this important publication on Oregon Mining has not been available for nearly seventy five years. Included are details on the geology of this unique portion of North Eastern Oregon. This is the first part of a two book series on the area. Accompanying the text are rare photographs and historic maps.8.5" X 11", 94 ppgs. Retail Price: $9.99

Field Identification of Minerals For Oregon Prospectors - Originally published in 1940, this important publication on Oregon Mining has not been available for nearly seventy five years. Included in this volume is an easy system for testing and identifying a wide range of minerals that might be found by prospectors, geologists and rockhounds in the State of Oregon, as well as in other locales. Topics include how to put together your own field testing kit and how to conduct rudimentary tests in the field. This volume is written in a clear and concise way to make it useful even for beginners. 8.5" X 11", 158 ppgs. Retail Price: $14.99

The Bohemia Mining District of Oregon - Originally published in 1900, this important publication on Oregon Mining has not been available for over a century. Included in this volume are important insights into the famous Bohemia Mining District of Oregon, including the histories and locations of important gold mines in the area such as the Ophir Mine, Clarence, Acturas, Peek-a-boo, White Swan, Combination Mine, the Musick Mine, The California, White Ghost, The Mystery, Wall Street, Vesuvius, Story, Lizzie Bullock, Delta, Elsie Dora, Golden Slipper, Broadway, Champion Mine, Knott, Noonday, Helena, White Wings, Riverside and others. Also included are notes on the nearby Blue River Mining District. 8.5" X 11", 58 ppgs. Retail Price: $9.99

The Gold Fields of Eastern Oregon - Unavailable since 1900, this publication was originally compiled by the Baker City Chamber of Commerce Offering important insights into the gold mining history of Eastern Oregon, "The Gold Fields of Eastern Oregon" sheds a rare light on many of the gold mines that were operating at the turn of the 19th Century in Baker County and Grant County in North Eastern Oregon. Some of the areas featured include the Cable Cove District, Baisely-Elhorn, Granite, Red Boy, Bonanza, Susanville, Sparta, Virtue, Vaughn, Sumpter, Burnt River, Rye Valley and other mining districts. Included is basic information on not only many gold mines that are well known to those interested in Eastern Oregon mining history, but also many mines and prospects which have been mostly lost to the passage of time. Accompanying are numerous rare photos 8.5" X 11", 78 ppgs. **Retail Price: $10.99**

Gold Mining in Eastern Oregon - Originally published in 1938, this important publication on Oregon Mining has not been available for over a century. Included in this volume are important insights into the famous mining districts of Eastern Oregon during the late 1930's. Particular attention is given to those gold mines with milling and concentrating facilities in the Greenhorn, Red Boy, Alamo, Bonanza, Granite, Cable Cove, Cracker Creek, Virtue, Keating, Medical Springs, Sanger, Sparta, Chicken Creek, Mormon Basin, Connor Creek, Cornucopia and the Bull Run Mining Districts. Some of the mines featured include the Ben Harrison, North Pole-Columbia, Highland Maxwell, Baisley-Elkhorn, White Swan, Balm Creek, Twin Baby, Gem of Sparta, New Deal, Gleason, Gifford-Johnson, Cornucopia, Record, Bull Run, Orion and others. Of particular interest are the mill flow sheets and descriptions of milling operations of these mines. 8.5" X 11", 68 ppgs. **Retail Price: $8.99**

The Gold Belt of the Blue Mountains of Oregon - Originally published in 1901, this important publication on Oregon Mining has not been available for over a century. Included in this volume are rare insights into the gold deposits of the Blue Mountains of North East Oregon, including the history of their early discovery and early production. Extensive details are offered on this important mining area's mineralogy and economic geology, as well as insights into nearby gold placers, silver deposits and copper deposits. Featured are the Elkhorn and Rock Creek mining districts, the Pocahontas district, Auburn and Minersville districts, Sumpter and Cracker Creek, Cable Cove, the Camp Carson district, Granite, Alamo, Greenhorn, Robinsonville, the Upper Burnt River Valley and Bonanza districts, Susanville, Quartzburg, Canyon Creek, Virtue, the Copper Butte district, the North Powder River, Sparta, Eagle Creek, Cornucopia, Pine Creek, Lower Powder River, the Upper Snake River Canyon, Rye Valley, Lower Burnt River Valley, Mormon Basin, the Malheur and Clarks Creek districts, Sutton Creek and others. Of particular interest are important details on numerous gold mines and prospects in these mining districts, including their locations, histories, geology and other important information, as well as information on silver, copper and fire opal deposits. 8.5" X 11", 250 ppgs. **Retail Price: $24.99**

Mining in the Cascades Range of Oregon - Originally published in 1938, this important publication on Oregon Mining has not been available for over seventy five years. Included in this volume are rare insights into the gold mines and other types of metal mines in the Cascades Mountain Range of Oregon. Some of the important mining areas covered include the famous Bohemia Mining District, the North Santiam Mining District, Quartzville Mining District, Blue River Mining District, Fall Creek Mining District, Oakridge District, Zinc District, Buzzard-Al Sarena District, Grand Cove, Climax District and Barron Mining District. Of particular interest are important details on over 100 mines and prospects in these mining districts, including their locations, histories, geology and other important information. 8.5" X 11", 170 ppgs. Retail Price: $14.99

Beach Gold Placers of the Oregon Coast - Originally published in 1934, this important publication on Oregon Mining has not been available for over 80 years. Included in this volume are rare insights into the beach gold deposits of the State of Oregon, including their locations, occurance, composition and geology. Of particular interest is information on placer platinum in Oregon's rich beach deposits. Also included are the locations and other information on some famous Oregon beach mines, including the Pioneer, Eagle, Chickamin, Iowa and beach placer mines north of the mouth of the Rogue River. 8.5" X 11", 60 ppgs. Retail Price: $8.99

Mineralogical Composition of the Sands of the Oregon Coast: From Coos Bay to the Columbia - Published in 1945, he text features hard to find information on the composition of the gold bearing black sands of the South West Oregon Coast, offering a unique insight to prospectors in search of Oregon's legendary beach gold. 104 ppgs, $9.99

Manganese Mining in Oregon - First released in 1942 and now out of print, this special reprint edition of "Manganese in Oregon" was originally published by the Oregon Department of Geology and Mineral Industries. The text features hard to find information on the mining of Manganese in Oregon, including details and maps of Oregon manganese mines and prospects. 108 ppgs, 9.99

Medford Oregon As A Mining Center - Written in 1912, this hard to find publication includes valuable insights into the mining history of South West Oregon. This small book contains interesting information on the gold, copper and mining industry in Southern Oregon as it existed just prior to World War One, shedding light on some of the important mines in the area. Included are rare photographs and vintage advertising of the day. 80 ppgs, 9.99

Mineral Resources of Curry County Oregon - First released in 1977 and now out of print, this special reprint edition of "Geology, Mineral Resources and Rock Materials of Curry County, Oregon" was originally published in cooperation of Curry County, Oregon and the Oregon Department of Geology and Mineral Industries. The text features hard to find information on not only the mining of gold and other metals in Curry County, but also aggregate mining in the area. 102 ppgs, 11.99

Origin of the Gold Bearing Black Sands of the Coast of South West Oregon - First released in 1943 and now out of print, this special reprint edition of "The Origin of the Black Sands of the South West Oregon Coast" was originally published by the Oregon Department of Geology and Mineral Industries. The text features hard to find information on the origin of the gold bearing black sands of the South West Oregon Coast, offering a unique insight to prospectors in search of Oregon's legendary beach gold. 52 ppgs, 8.99

South West Oregon Mining - Leading mining historian Kerby Jackson introduces us to six classic small mining publications on the Gold Mining Industry in Southern Oregon. This small book consists of a compilation of USGS J.S. Diller's "Mines of the Riddles Quadrangle", "The Rogue River Valley Coal Fields" and "Mineral Resources of the Grants Pass Quadrangle", the Grants Pass Commercial Club's rare publication "Mining in Josephine County, Oregon" and the USGS publication "The Distribution of Placer Gold in the Sixes River, South West Oregon". Also included is F.W. Libbey's legendary article on the Southern Oregon Mining Industry, "Lest We Forget", which appeared in the publication of the Oregon State Department of Geology and Mineral Industries in the early 1960's. This compilation offers a unique perspective on mining in South West Oregon and includes considerable information on mines in Josephine, Jackson and Coos Counties. 142 ppgs, 14.99

Geology and Mineral Resources of the Gasquet Quadrangle of California-Oregon - First published in 1953, it has been unavailable for over a century and sheds important light on the geological features and mineral resources of this portion of Northern California and Southern Oregon. 80 ppgs, 9.99

Idaho Mining Books

Gold in Idaho - Unavailable since the 1940's, this publication was originally compiled by the Idaho Bureau of Mines and includes details on gold mining in Idaho. Included is not only raw data on gold production in Idaho, but also valuable insight into where gold may be found in Idaho, as well as practical information on the gold bearing rocks and other geological features that will assist those looking for placer and lode gold in the State of Idaho. This volume also includes thirteen gold maps that greatly enhance the practical usability of the information contained in this small book detailing where to find gold in Idaho. **8.5" X 11", 72 ppgs. Retail Price: $9.99**

Geology of the Couer D'Alene Mining District of Idaho - Unavailable since 1961, this publication was originally compiled by the Idaho Bureau of Mines and Geology and includes details on the mining of gold, silver and other minerals in the famous Coeur D'Alene Mining District in Northern Idaho. Included are details on the early history of the Coeur D'Alene Mining District, local tectonic settings, ore deposit features, information on the mineral belts of the Osburn Fault, as well as detailed information on the famous Bunker Hill Mine, the Dayrock Mine, Galena Mine, Lucky Friday Mine and the infamous Sunshine Mine. This volume also includes sixteen hard to find maps. **8.5" X 11", 70 ppgs. Retail Price: $9.99**

The Gold Camps and Silver Cities of Idaho - Originally published in 1963, this important publication on Idaho Mining has not been available for nearly fifty years. Included are rare insights into the history of Idaho's Gold Rush, as well as the mad craze for silver in the Idaho Panhandle. Documented in fine detail are the early mining excitements at Boise Basin, at South Boise, in the Owyhees, at Deadwood, Long Valley, Stanley Basin and Robinson Bar, at Atlanta, on the famous Boise River, Volcano, Little Smokey, Banner, Boise Ridge, Hailey, Leesburg, Lemhi, Pearl, at South Mountain, Shoup and Ulysses, Yellow Jacket and Loon Creek. The story follows with the appearance of Chinese miners at the new mining camps on the Snake River, Black Pine, Yankee Fork, Bay Horse, Clayton, Heath, Seven Devils, Gibbonsville, Vienna and Sawtooth City. Also included are special sections on the Idaho Lead and Silver mines of the late 1800's, as well as the mining discoveries of the early 1900's that paved the way for Idaho's modern mining and mineral industry. Lavishly illustrated with rare historic photos, this volume provides a one of a kind documentary into Idaho's mining history that is sure to be enjoyed by not only modern miners and prospectors who still scour the hills in search of nature's treasures, but also those enjoy history and tromping through overgrown ghost towns and long abandoned mining camps. **8.5" X 11", 186 ppgs. Retail Price: $14.99**

Ore Deposits and Mining in North Western Custer County Idaho - Unavailable since 1913, this important publication was originally published by the Us Department of the Interior and has been unavailable for a century. Included are fine details on the geology, geography, gold placers and gold and silver bearing quartz veins of the mining region of North West Custer County, Idaho. Of particular interest is a rare look at the mines and prospects of the region, including those such as the Ramshorn Mine, SkyLark, Riverview, Excelsior, Beardsley, Pacific, Hoosier, Silver Brick, Forest Rose and dozens of others in the Bay Horse Mining District. Also covered are the mines of the Yankee Fork District such as the Lucky Boy, Badger, Black, Enterprise, Charles Dickens, Morrison, Golden Sunbeam, Montana, Golden Gate and others, as well as those in the Loon Mining District. **8.5" X 11", 126 ppgs. Retail Price: $12.99**

Gold Rush To Idaho - Unavailable since 1963, this important publication was originally published by the Idaho Bureau of Mines and has been unavailable for 50 years. "Gold Rush To Idaho" revisits the earliest years of the discovery of gold in Idaho Territory and introduces us to the conditions that the pioneer gold seekers met when they blazed a trail through the wilderness of Idaho's mountains and discovered the precious yellow metal at Oro Fino and Pierce. Subsequent rushes followed at places like Elk City, Newsome, Clearwater Station, Florence, Warrens and elsewhere. Of particular interest is a rare look at the hardships that the first miners in Idaho met with during their day to day existences and their attempts to bring law and order to their mining camps. **8.5" X 11", 88 ppgs. Retail Price: $9.99**

The Geology and Mines of Northern Idaho and North Western Montana - Unavailable since 1909, this important publication was originally published by the Us Department of the Interior and has been unavailable for a century. Included are fine details on the geology and geography of the mining regions of Northern Idaho and North Western Montana. Of particular interest is a rare look at the mines and prospects of the region, including those in the Pine Creek Mining District, Lake Pend Oreille district, Troy Mining District, Sylvanite District, Cabinet Mining District, Prospect Mining District and the Missoula Valley. Some of the mines featured include the Iron Mountain, Silver Butte, Snowshoe, Grouse Mountain Mine and others. **8.5" X 11", 142 ppgs. Retail Price: $12.99**

Mining in the Alturas Quadrangle of Blaine County Idaho - Unavailable since 1922, this important publication was originally published by the Idaho Bureau of Mines and has been unavailable for ninety years. Topics include the geology, rock formations and the formation of ore deposits in this important mining area of Idaho. Of particular focus is information on the local geology, quartz veins and ore deposits of this portion of Idaho. Included are hard to find details, including the descriptions and locations of numerous gold and silver mines in the area including the Silver King, Pilgrim, Columbia, Lone Jack, Sunbeam, Pride of the West, Lucky Boy, Scotia, Atlanta, Beaver-Bidwell and others mines and prospects. **8.5" X 11", 56 ppgs. Retail Price: $8.99**

Mining in Lemhi County Idaho - Originally published in 1913, this important book on Idaho Mining has not been available to miners for over a century. Included are rare insights into hundreds of gold, silver, copper and other mines in this famous Idaho mining area. Details include the locations, geology, history, production and other facts of the mines of this region, not only gold and silver hardrock mines, but also gold placer mines, lead-silver deposits, copper mines, cobalt-nickel deposits, tungsten and tin mines . It is lavishly illustrated with hard to find photos of the period and rare mining maps. Some of the vicinities featured include the Nicholia Mining District, Spring Mountain District, Texas District, Blue Wing District, Junction District, McDevitt District, Pratt Creek, Eldorado District, Kirtley Creek, Carmen Creek, Gibbonsville, Indian Creek, Mineral Hill District, Mackinaw, Eureka District, Blackbird District, YellowJacket District, Gravel Range District, Junction District, Parker Mountain and other mining districts. 8.5" X 11", 226 ppgs. Retail Price: $19.99

Mining in Shoshone County Idaho - First published in 1923, it has been unavailable for over a century and sheds important light on the mining history of Shoshone County, Idaho. Some of the topics include the history of mining in Shoshone County, a look at the local geology and ore characteristics of lead-silver deposits, zinc deposits, copper, antimony, gold and other minerals. Also included are insights into the history, production, characteristics and locations of numerous mines in the area. 198 ppgs, 15.99

Utah Mining Books

Fluorite in Utah - Unavailable since 1954, this publication was originally compiled by the USGS, State of Utah and U.S. Atomic Energy Commission and details the mining of fluorspar, also known as fluorite in the State of Utah. Included are details on the geology and history of fluorspar (fluorite) mining in Utah, including details on where this unique gem mineral may be found in the State of Utah. 8.5" X 11", 60 ppgs. Retail Price: $8.99

The Gold Hill Mining District of Utah - First published in 1935, it has been unavailable since those days and sheds important light on the mines, history and geology of Utah's Gold Hill Mining District. Included are rare insights into this important mining area, including the locations, histories and details of numerous mines. This volume is well illustrated with geological diagrams, as well as hard to find maps of some of the most important mines in this district. 202 ppgs., 19.99

The Mines, Miners and Minerals of Utah - First published in 1896, it has been unavailable since those days and sheds important light on the early mines and miners of Pioneer Utah, as well as the minerals which they won from the earth by laborious hard physical labor and sheer determination. Included are rare insights into the early mining history of Utah, as well details on hundreds of gold, silver and copper mines. 376 ppgs., 24.99

California Mining Books

The Tertiary Gravels of the Sierra Nevada of California - Mining historian Kerby Jackson introduces us to a classic mining work by Waldemar Lindgren in this important re-issue of The Tertiary Gravels of the Sierra Nevada of California. Unavailable since 1911, this publication includes details on the gold bearing ancient river channels of the famous Sierra Nevada region of California. 8.5" X 11", 282 ppgs. Retail Price: $19.99

The Mother Lode Mining Region of California - Unavailable since 1900, this publication includes details on the gold mines of California's famous Mother Lode gold mining area. Included are details on the geology, history and important gold mines of the region, as well as insights into historic mining methods, mine timbering, mining machinery, mining bell signals and other details on how these mines operated. Also included are insights into the gold mines of the California Mother Lode that were in operation during the first sixty years of California's mining history. 8.5" X 11", 176 ppgs. Retail Price: $14.99

Lode Gold of the Klamath Mountains of Northern California and South West Oregon - Unavailable since 1971, this publication was originally compiled by Preston E. Hotz and includes details on the lode mining districts of Oregon and California's Klamath Mountains. Included are details on the geology, history and important lode mines of the French Gulch, Deadwood, Whiskeytown, Shasta, Redding, Muletown, South Fork, Old Diggings, Dog Creek (Delta), Bully Choop (Indian Creek), Harrison Gulch, Hayfork, Minersville, Trinity Center, Canyon Creek, East Fork, New River, Denny, Liberty (Black Bear), Cecilville, Callahan, Yreka, Fort Jones and Happy Camp mining districts in California, as well as the Ashland, Rogue River, Applegate, Illinois River, Takilma, Greenback, Galice, Silver Peak, Myrtle Creek and Mule Creek districts of South Western Oregon. Also included are insights into the mineralization and other characteristics of this important mining region. 8.5" X 11", 100 ppgs. Retail Price: $10.99

Mines and Mineral Resources of Shasta County, Siskiyou County, Trinity County: California - Unavailable since 1915, this publication was originally compiled by the California State Mining Bureau and includes details on the gold mines of this area of Northern California. Also included are insights into the mineralization and other characteristics of this important mining region, as well as the location of historic gold mines. 8.5" X 11", 204 ppgs. Retail Price: $19.99

Geology of the Yreka Quadrangle, Siskiyou County, California - Unavailable since 1977, this publication was originally compiled by Preston E. Hotz and includes details on the geology of the Yreka Quadrangle of Siskiyou County, California. Also included are insights into the mineralization and other characteristics of this important mining region. **8.5" X 11", 78 ppgs. Retail Price: $7.99**

Mines of San Diego and Imperial Counties, California - Originally published in 1914, this important publication on California Mining has not been available for a century. This publication includes important information on the early gold mines of San Diego and Imperial County, which were some of the first gold fields mined in California by early Spanish and Mexican miners before the 49ers came on the scene. Included are not only details on early mining methods in the area, production statistics and geological information, but also the location of the early gold mines that helped make California "The Golden State". Also included are details on the mining of other minerals such as silver, lead, zinc, manganese, tungsten, vanadium, asbestos, barite, borax, cement, clay, dolomite, fluospar, gem stones, graphite, marble, salines, petroleum, stronium, talc and others. **8.5" X 11", 116 ppgs. Retail Price: $12.99**

Mines of Sierra County, California - Unavailable since 1920, this publication was originally compiled by the California State Mining Bureau and includes details on the gold mines of Sierra County, California. Also included are insights into the mineralization and other characteristics of this important mining region, as well as the location of historic gold mines. **8.5" X 11", 156 ppgs. Retail Price: $19.99**

Mines of Plumas County, California - Unavailable since 1918, this publication was originally compiled by the California State Mining Bureau and includes details on the gold mines of Plumas County, California. Also included are insights into the mineralization and other characteristics of this important mining region, as well as the location of historic gold mines. **8.5" X 11", 200 ppgs. Retail Price: $19.99**

Mines of El Dorado, Placer, Sacramento and Yuba Counties, California - Originally published in 1917, this important publication on California Mining has not been available for nearly a century. This publication includes important information on the early gold mines of El Dorado County, Placer County, Sacramento County and Yuba County, which were some of the first gold fields mined by the Forty-Niners during the California Gold Rush. Included are not only details on early mining methods in the area, production statistics and geological information, but also the location of the early gold mines that helped make California "The Golden State". Also included are insights into the early mining of chrome, copper and other minerals in this important mining area. **8.5" X 11", 204 ppgs. Retail Price: $19.99**

Mines of Los Angeles, Orange and Riverside Counties, California - Originally published in 1917, this important publication on California Mining has not been available for nearly a century. This publication includes important information on the early gold mines of Los Angeles County, Orange County and Riverside County, which were some of the first gold fields mined in California by early Spanish and Mexican miners before the 49ers came on the scene. Included are not only details on early mining methods in the area, production statistics and geological information, but also the location of the early gold mines that helped make California "The Golden State". **8.5" X 11", 146 ppgs. Retail Price: $12.99**

Mines of San Bernadino and Tulare Counties, California - Originally published in 1917, this important publication on California Mining has not been available for nearly a century. This publication includes important information on the early gold mines of San Bernadino and Tulare County, which were some of the first gold fields mined in California by early Spanish and Mexican miners before the 49ers came on the scene. Included are not only details on early mining methods in the area, production statistics and geological information, but also the location of the early gold mines that helped make California "The Golden State". Also included are details on the mining of other minerals such as copper, iron, lead, zinc, manganese, tungsten, vanadium, asbestos, barite, borax, cement, clay, dolomite, fluospar, gem stones, graphite, marble, salines, petroleum, stronium, talc and others. **8.5" X 11", 200 ppgs. Retail Price: $19.99**

Chromite Mining in The Klamath Mountains of California and Oregon - Unavailable since 1919, this publication was originally compiled by J.S. Diller of the United States Department of Geological Survey and includes details on the chromite mines of this area of Northern California and Southern Oregon. Also included are insights into the mineralization and other characteristics of this important mining region, as well as the location of historic mines. Also included are insights into chromite mining in Eastern Oregon and Montana. **8.5" X 11", 98 ppgs. Retail Price: $9.99**

Mines and Mining in Amador, Calaveras and Tuolumne Counties, California - Unavailable since 1915, this publication was originally compiled by William Tucker and includes details on the mines and mineral resources of this important California mining area. Included are details on the geology, history and important gold mines of the region, as well as insights into other local mineral resources such as asbestos, clay, copper, talc, limestone and others. Also included are insights into the mineralization and other characteristics of this important portion of California's Mother Lode mining region. **8.5" X 11", 198 ppgs. Retail Price: $14.99**

The Cerro Gordo Mining District of Inyo County California - Unavailable since 1963, this publication was originally compiled by the United States Department of Interior. Included are insights into the mineralization and other characteristics of this important mining region of Southern California. Topics include the mining of gold and silver in this important mining district in Inyo County, California, including details on the history, production and locations of the Cerro Gordo Mine, the Morning Star Mine, Estelle Tunnel, Charles Lease Tunnel, Ignacio, Hart, Crosscut Tunnel, Sunset, Upper Newtown, Newtown, Ella, Perseverance, Newsboy, Belmont and other silver and gold mines in the Cerro Gordo Mining District. This volume also includes important insights into the fossil record, geologic formations, faults and other aspects of economic geology in this California mining district. 8.5" X 11", 104 ppgs. Retail Price: $10.99

Mining in Butte, Lassen, Modoc, Sutter and Tehama Counties of California - Unavailable since 1917, this publication was originally compiled by the United States Department of Interior. Included are insights into the mineralization and other characteristics of this important mining region of California. Topics include the mining of asbestos, chromite, gold, diamonds and manganese in Butte County, the mining of gold and copper in the Hayden Hill and Diamond Mountain mining districts of Lassen County, the mining of coal, salt, copper and gold in the High Grade and Winters mining districts of Modoc County, gold mining in Sutter County and the mining of gold, chromite, manganese and copper in Tehama County. This volume also includes the production records and locations of numerous mines in this important mining region. 8.5" X 11", 114 ppgs. Retail Price: $11.99

Mines of Trinity County California - Originally published in 1965, this important publication on California Mining has not been available for nearly fifty years. This publication includes important information on mines and mining in Trinity County, California, as well insights into the mineralization and geology of this important mining area in Northern California. Included are extensive details on hardrock and placer gold mines and prospects, including charts showing the locations of these historic mines.. 8.5" X 11", 144 ppgs. Retail Price: $12.99

Mines of Kern County California - Originally published in 1962, this important publication on California Mining has not been available for nearly fifty years. This publication includes important information on mines and mining in Kern County, California, as well insights into the mineralization and geology of this important mining area in California. Included are extensive details on hardrock and placer gold mines and prospects, including charts showing the locations of these historic mines. 8.5" X 11", 398 ppgs. Retail Price: $24.99

Mines of Calaveras County California - Originally published in 1962, this important publication on California Mining has not been available for nearly fifty years. This publication includes important information on mines and mining in Calaveras County, California, as well insights into the mineralization and geology of this important mining area in Northern California. Included are extensive details on hardrock and placer gold mines and prospects, including charts showing the locations of these historic mines. 8.5" X 11", 236 ppgs. Retail Price: $19.99

Lode Gold Mining in Grass Valley California - Unavailable since 1940, this publication was originally compiled by the United States Department of Interior. Included are insights into the gold mineralization and other characteristics of this important mining region of Nevada County, California. This volume also includes important insights into the geologic formations, faults and other aspects of economic geology in this California mining district. Of particular interest are the fine details on many hardrock gold mines in the area, including their locations, histories, development and mineralization. Some of the mines featured include the Gold Hill Mine, Massachusetts Hill, Boundary, Peabody, Golden Center, North Star, Omaha, Lone Jack, Homeward Bound, Hartery, Wisconsin, Allison Ranch, Phoenix, Kate Hayes, W.Y.O.D., Empire, Rich Hill, Daisy Hill, Orleans, Sultana, Centennial, Conlin, Ben Franklin, Crown Point and many others. 8.5" X 11", 148 ppgs. Retail Price: $12.99

Lode Mining in the Alleghany District of Sierra County California - Unavailable since 1913, this publication was originally compiled by the United States Department of Interior. Included are insights into the mineralization and other characteristics of this important mining region of Sierra County. Included are details on the history, production and locations of numerous hardrock gold mines in this famous California area, including the Tightner Mine, Minnie D., Osceola, Eldorado, Twenty One, Sherman, Kenton, Oriental, Rainbow, Plumbago, Irelan, Gold Canyon, North Fork, Federal, Kate Hardy and others. This volume also includes important insights into the fossil record, geologic formations, faults and other aspects of economic geology in this California mining district. 8.5" X 11", 48 ppgs. Retail Price: $7.99

Six Months In The Gold Mines During The California Gold Rush - Unavailable since 1850, this important work is a first hand account of one "49'ers" personal experience during the great California Gold Rush, shedding important light on one of the most exciting periods in the history of not only California, but also the world. Compiled from journals written between 1847 and 1849 by E. Gould Buffum, a native of New York, "Six Months In The Gold Mines During The California Gold Rush" offers a rare look into the day to day lives of the people who came to California to work in her gold mines when the state was still a great frontier. 8.5" X 11", 290 ppgs. Retail Price: $19.99

<u>Quartz Mines of the Grass Valley Mining District of California</u> - Unavailable since 1867, this important publication has not been available since those days. This rare publication offers a short dissertation on the early hardrock mines in this important mining district in the California Mother Lode region between the 1850's and 1860's. Also included are hard to find details on the mineralization and locations of these mines, as well as how they were operated in those day. 8.5" X 11", 44 ppgs. Retail Price: $8.99

<u>Gold Rush on the Feather River</u> - First published in 1924, this short publication by G.C. Mansfield sheds important light on the early history of gold mining on the Feather River. Included are rare insights into the first decade of gold mining and the early mining camps of the Feather River during the 1850's. 64 ppgs., 9.99

<u>The Bodie Mining District of California</u> - First published in 1986, it has been unavailable since those days and sheds important light on this famous mining area. Included are the history, characteristics and locations of numerous old mines around the ghost town of Bodie.
64 ppgs, 8.99

<u>Geology and Mineral Resources of the Gasquet Quadrangle of California-Oregon</u> - First published in 1953, it has been unavailable for over a century and sheds important light on the geological features and mineral resources of this portion of Northern California and Southern Oregon.
80 ppgs, 9.99

Alaska Mining Books

<u>Ore Deposits of the Willow Creek Mining District, Alaska</u> - Unavailable since 1954, this hard to find publication includes valuable insights into the Willow Creek Mining District near Hatcher Pass in Alaska. The publication includes insights into the history, geology and locations of the well known mines in the area, including the Gold Cord, Independence, Fern, Mabel, Lonesome, Snowbird, Schroff-O'Neil, High Grade, Marion Twin, Thorpe, Webfoot, Kelly-Willow, Lane, Holland and others. 8.5" X 11", 96 ppgs. Retail Price: $9.99

<u>The Juneau Gold Belt of Alaska</u> - Unavailable since 1906, this hard to find publication includes valuable insights into the gold mines around Juneau, Alaska. The publication includes important details into the history, geology and locations of the well known gold mines and prospects in the area, including those around Windham Bay, Holkham Bay, Port Snettisham, on Grindstone and Rhine Creeks, Gold Creek, Douglas Island, Salmon Creek, Lemon Creek, Nugget Creek, from the Mendenhall River to Berners Bay, McGinnis Creek, Montana Creek, Peterson Creek, Windfall Creek, the Eagle River, Yankee Basin, Yankee Curve, Kowee Creek and elsewhere. Not only are gold placer mines included, but also hardrock gold mines. 8.5" X 11", 224 ppgs. Retail Price: $19.99

<u>Mining in the Jumbo Basin of Alaska</u> - Unavailable since 1953, this hard to find publication includes valuable insights into the mines and geology of the Jumbo Basin. The publication includes important details into the history, geology and locations of the well known gold mines and prospects in the famous Jumbo Basin Mining Region of Alaska.
72 ppgs, 9.99

<u>The Rampart Placer Gold Region of Alaska</u> - Unavailable since 1906, this hard to find publication includes valuable insights into the placer gold mines of the Rampart Mining Region. The publication includes important details into the history, geology and locations of the well known gold mines and prospects in the famous Rampart Mining Region of Alaska.
78 ppgs, 10.99

Arizona Mining Books

<u>Mines and Mining in Northern Yuma County Arizona</u> - Originally published in 1911, this important publication on Arizona Mining has not been available for over a hundred years. Included are rare insights into the gold, silver, copper and quicksilver mines of Yuma County, Arizona together with hard to find maps and photographs. Some of the mines and mining districts featured include the Planet Copper Mine, Mineral Hill, the Clara Consolidated Mine, Viati Mine, Copper Basin prospect, Bowman Mine, Quartz King, Billy Mack, Carnation, the Wardwell and Osbourne, Valensuella Copper, the Mariquita, Colonial Mine, the French American, the New York-Plomosa, Guadalupe, Lead Camp, Mudersbach Copper Camp, Yellow Bird, the Arizona Northern (Salome Strike), Bonanza (Harqua Hala), Golden Eagle, Hercules, Socorro and others. 8.5" X 11", 144 ppgs. Retail Price: $11.99

<u>The Aravaipa and Stanley Mining Districts of Graham County Arizona</u> - Originally published in 1925, this important publication on Arizona Mining has not been available for nearly ninety years. Included are rare insights into the gold and silver mines of these two important mining districts, together with hard to find maps. 8.5" X 11", 140 ppgs. Retail Price: $11.99

Gold in the Gold Basin and Lost Basin Mining Districts of Mohave County, Arizona - This volume contains rare insights into the geology and gold mineralization of the Gold Basin and Lost Basin Mining Districts of Mohave County, Arizona that will be of benefit to miners and prospectors. Also included is a significant body of information on the gold mines and prospects of this portion of Arizona. This volume is lavishly illustrated with rare photos and mining maps. 8.5" X 11", 188 ppgs. Retail Price: $19.99

Mines of the Jerome and Bradshaw Mountains of Arizona - This important publication on Arizona Mining has not been available for ninety years. This volume contains rare insights into the geology and ore deposits of the Jerome and Bradshaw Mountains of Arizona that will be of benefit to miners and prospectors who work those areas. Included is a significant body of information on the mines and prospects of the Verde, Black Hills, Cherry Creek, Prescott, Walker, Groom Creek, Hassayampa, Bigbug, Turkey Creek, Agua Fria, Black Canyon, Peck, Tiger, Pine Grove, Bradshaw, Tintop, Humbug and Castle Creek Mining Districts. This volume is lavishly illustrated with rare photos and mining maps. 8.5" X 11", 218 ppgs. Retail Price: $19.99

The Ajo Mining District of Pima County Arizona - This important publication on Arizona Mining has not been available for nearly seventy years. This volume contains rare insights into the geology and mineralization of the Ajo Mining District in Pima County, Arizona and in particular the famous New Cornelia Mine. 8.5" X 11", 126 ppgs. Retail Price: $11.99

Mining in the Santa Rita and Patagonia Mountains of Arizona - Originally published in 1915, this important publication on Arizona Mining has not been available for nearly a century. Included are rare insights into hundreds of gold, silver, copper and other mines in this famous Arizona mining area. Details include the locations, geology, history, production and other facts of the mines of this region. 8.5" X 11", 394 ppgs. Retail Price: $24.99

Mining in the Bisbee Quadrangle of Arizona - Originally published in 1906, this important publication on Arizona Mining has not been available for nearly a century. Included are rare insights into hundreds of gold, silver, copper and other mines in this famous Arizona mining area. Details include the locations, geology, history, production and other facts of the mines of this important mining region. 8.5" X 11", 188 ppgs. Retail Price: $14.99

Placer Gold Mining in Arizona - Unavailable since 1922, this hard to find publication includes valuable insights into the placer gold mines of the Arizona. Originally released as "Placer Gold of Arizona", despite its small size, this publication includes important details into the history, geology and locations of the well known placer gold mines and prospects in the State of Arizona. 48 ppgs, 8.99

Gold and Copper Mining near Payson, Arizona - Written in 1915, this hard to find publication includes valuable insights into the gold and copper mining industry of Arizona. Highlighted here are the gold and copper mines near Payson, Arizona. 68 ppgs, 8.99

Lode Gold Mining in Arizona - Unavailable since 1934, this hard to find publication, originally released as "Arizona Lode Gold Mines and Gold Mining" includes valuable insights into the gold mining industry of Arizona. Included are valuable insights into over 150 hardrock gold mines in over 30 different mining districts in Arizona. 278 ppgs, 21.99

Mining in the Dragoon Quadrangle of Cochise County, Arizona - Unavailable since 1964, this hard to find publication includes valuable insights into the mines of the Dragoon Quadrangle Mining Region. The publication includes important details into the history, geology and locations of the well known mines and prospects in this famous mining region of Arizona. 224 ppgs., 19.99

Directory of Operating Mines in Arizona in 1915 - Unavailable since 1916, this hard to find publication includes valuable insights into the mines of Arizona. This small publication includes a complete list of the mines that were operating in the State of Arizona during 1915 and includes details such as general location, owners and some basic facts about each mining operation. 52 ppgs. 8.99

Arizona Ore Deposits - Unavailable since 1938, this hard to find publication includes valuable insights into some ore deposits of Arizona. Included are valuable insights into the formation and characteristics of valuable ore deposits in the Jerome, Miami, Inspiration, Clifton, Morenci, Ray, Ajo, Eureka, Tombstone and Magma mining districts. Included are details into some of the major gold, silver and copper mines of these important Arizona mining areas. 160 ppgs, 14.99

Montana Mining Books

A History of Butte Montana: The World's Greatest Mining Camp - First published in 1900 by H.C. Freeman, this important publication sheds a bright light on one of the most important mining areas in the history of The West. Together with his insights, as well as rare photographs of the periods, Harry Freeman describes Butte and its vicinity from its early beginnings, right up to its flush years when copper flowed from its mines like a river. At the time of publication, Butte, Montana was known worldwide as "The Richest Mining Spot On Earth" and produced not only vast amounts of copper, but also silver, gold and other metals from its mines. Freeman illustrates, with great detail, the most important mines in the vicinity of Butte, providing rare details on their owners, their history and most importantly, how the mines operated and how their treasures were extracted. Of particular interest are the dozens of rare photographs that depict mines such as the famous Anaconda, the Silver Bow, the Smoke House, Moose, Paulin, Buffalo, Little Minah, the Mountain Consolidated, West Greyrock, Cora, the Green Mountain, Diamond, Bell, Parnell, the Neversweat, Nipper, Original and many others. **8.5" X 11", 142 ppgs. Retail Price: $12.99**

The Butte Mining District of Montana - This important publication on Montana Mining has not been available for over a century. Included are rare insights into the gold, copper and silver mines of Butte, Montana together with hard to find maps and photographs. Some of the topics include the early history of gold, silver and copper mining in the Butte area, insight into the geology of its mining areas, the local distribution of gold, silver and copper ores, as well their composition and how to identify them. Also included are detailed facts about the mines in the Butte Mining District, including the famous Anaconda Mine, Gagnon, Parrot, Blue Vein, Moscow, Poulin, Stella, Buffalo, Green Mountain, Wake Up Jim, the Diamond-Bell Group, Mountain Consolidated, East Greyrock, West Greyrock, Snowball, Corra, Speculator, Adirondack, Miners Union, the Jessie-Edith May Group, Otisco, Iduna, Colorado, Lizzie, Cambers, Anderson, Hesperus, Preferencia and dozens of others. **8.5" X 11", 298 ppgs. Retail Price: $24.99**

Mines of the Helena Mining Region of Montana - This important publication on Montana Mining has not been available for over a century. Included are rare insights into the gold, copper and silver mines of the vicinity of Helena, Montana, including the Marysville Mining District, Elliston Mining District, Rimini Mining District, Helena Mining District, Clancy Mining District, Wickes Mining District, Boulder and Basin Mining Districts and the Elkhorn Mining District. Some of the topics include the early history of gold, silver and copper mining in the Helena area, insight into the geology of its mining areas, the local distribution of gold, silver and copper ores, as well their composition and how to identify them. Also included are detailed facts, history, geology and locations of over one hundred gold, silver and copper mines in the area . **8.5" X 11", 162 ppgs, Retail Price: $14.99**

Mines and Geology of the Garnet Range of Montana - This important publication on Montana Mining has not been available for over a century. Included are rare insights into the gold, copper and silver mines of the vicinity of this important mining area of Montana. Some of the topics include the early history of gold, silver and copper mining in the Garnet Mountains, insight into the geology of its mining areas, the local distribution of gold, silver and copper ores, as well their composition and how to identify them. Also included are detailed facts, history, geology and locations of numerous gold, silver and copper mines in the area . **8.5" X 11", 100 ppgs, Retail Price: $11.99**

Mines and Geology of the Philipsburg Quadrangle of Montana - This important publication on Montana Mining has not been available for over a century. Included are rare insights into the gold, copper and silver mines of the vicinity of this important mining area of Montana. Some of the topics include the early history of gold, silver and copper mining in the Philipsburg Quadrangle, insight into the geology of its mining areas, the local distribution of gold, silver and copper ores, as well their composition and how to identify them. Also included are detailed facts, history, geology and locations of over one hundred gold, silver and copper mines in the area **8.5" X 11", 290 ppgs, Retail Price: $24.99**

Geology of the Marysville Mining District of Montana - Included are rare insights into the mining geology of the Marysville Mining District. Some of the topics include the early history of gold, silver and copper mining in the area, insight into the geology of its mining areas, the local distribution of gold, silver and copper ores, as well their composition and how to identify them. Also included are detailed facts, history, geology and locations of gold, silver and copper mines in the area **8.5" X 11", 198 ppgs, Retail Price: $19.99**

The Geology and Mines of Northern Idaho and North Western Montana- See listing under Idaho.

The History of Gold Dredging in Montana - Unavailable since 1916, this important publication was originally published by the Us Bureau of Mines and has been unavailable for a century. A century and more ago, giant dredging machines dug in Montana's rivers and creeks in search of illusive golden riches. First appearing in California in the 1850's, gold dredges finally reached their peak of development in Siberia and New Zealand before becoming popular again in the United States. This book offers a unique historical perspective on the gold dredges that once operated in Montana. This book on Montana mining history is lavishly illustrated with dozens of rare historic photos gold dredges that once operated in Montana, as well as hard to locate plans on how these dredges were designed. 120 ppgs., 11.99

Nevada Mining Books

The Bull Frog Mining District of Nevada - Unavailable since 1910, this publication was originally compiled by the United States Department of Interior. This volume also includes important insights into the geologic formations, faults and other aspects of economic geology in this Nevada mining district. Of particular interest are the fine details on many mines in the area, including their locations, histories, development and mineralization. Some of the mines featured include the National Bank Mine, Providence, Gibraltor, Tramps, Denver, Original Bullfrog, Gold Bar, Mayflower, Homestake-King and other mines and prospects. **8.5" X 11", 152 ppgs, Retail Price: $14.99**

History of the Comstock Lode - Unavailable since 1876, this publication was originally released by John Wiley & Sons. This volume also includes important insights into the famous Comstock Lode of Nevada that represented the first major silver discovery in the United States. During its spectacular run, the Comstock produced over 192 million ounces of silver and 8.2 million ounces of gold. Not only did the Comstock result in one of the largest mining rushes in history and yield immense fortunes for its owners, but it made important contributions to the development of the State of Nevada, as well as neighboring California. Included here are important details on not only the early development and history of the Comstock, but also rare early insight into its mines, ore and its geology.8.5" X 11", 244 ppgs, Retail Price: $19.99

The Pioche Mining District of Nevada - First published in 1932, it has been unavailable for over a century and sheds important light on the mining history of Nevada. Some of the topics include the history of mining in this district, as well as the characteristics of its mineral and ore deposits. Also included are insights into the history, production, characteristics and locations of numerous mines in the area. Some of the mines include the Combined Metals, Pioche, Ely Valley, No. 10, Poorman, Wide Awake, Alps, Prince, Virginia Louise, Half Moon, Abe Lincoln, Fairview, Bristol Silver, National, Vesuvius, Inman, Tempest, Hillside, Jackrabbit, Lucky Star, Fortuna, Mendha, Manhattan, Hamburg, Comet, Lyndon and others. 108 ppgs 10.99

The Yerington Mining District of Nevada - First published in 1932, it has been unavailable for over a century and sheds important light on the mining history of Nevada. Some of the topics include the history of mining in this district, as well as the characteristics of its mineral and ore deposits. Also included are insights into the history, production, characteristics and locations of numerous mines in the area. Some of the mines include the Bluestone, Mason Valley, Malachite, McConnell, Greenwood, Western Nevada, Ludwig, Douglas Hill, Casting Copper, Montana-Yerington, Empire, Jim Beatty, Terry and McFarland, Blue Jay and others. 92 ppgs, 10.99

The Genesis of the Ores of Tonopah Nevada - Unavailable since 1918, this hard to find publication includes valuable insights into the gold mines around Tonopah, Nevada. The publication includes important details into the geology of mines in the Tonopah Mining District of Nevada. 90 ppgs, 10.99

Mining Camps of Elko, Lander and Eureka Counties Nevada - Unavailable since 1910, this hard to find publication includes valuable insights into the mining camps of Elko, Lander and Eureka Counties, Nevada. The publication includes important details into the history of mines and mining in these three Nevada counties. 154 ppgs, 12.99

Ore Deposits of the Bullfrog Quadrangle - Unavailable since 1964 and released as "Geology of Bullfrog Quadrangle and Ore Deposits Related to Bullfrog Hills Caldera, Nye County, Nevada and Inyo County, California". The publication includes important details into the geology of mines in the Bullfrog Quadrangle of Nye County, Nevada and Inyo County, California. 52 ppgs, 9.99

Mining in Eureka County Nevada - Unavailable since 1879, this hard to find publication includes valuable insights into the early mining history off Eureka County, Nevada. The publication includes important details into the early history of the mines of Eureka County, as well as their development, production and how their ores were treated. Also included are details on the 1872 Mining Act, as well as the local rules, regulations and customs of the miners in Eureka County.134 ppgs, 12.99

Colorado Mining Books

Ores of The Leadville Mining District - Unavailable since 1926, this publication was originally compiled by the United States Department of Interior. This volume also includes important insights into the ores and mineralization of the Leadville Mining District in Colorado. Topics include historic ore prospecting methods, local geology, insights into ore veins and stockworks, the local trend and distribution of ore channels, reverse faults, shattered rock above replacement ore bodies, mineral enrichment in oxidized and sulphide zones and more. **8.5" X 11", 66 ppgs, Retail Price: $8.99**

Mining in Colorado - Unavailable since 1926, this publication was originally compiled by the United States Department of Interior. This volume also includes important insights into the mining history of Colorado from its early beginnings in the 1850's right up to the mid 1920's. Not only is Colorado's gold mining heritage included, but also its silver, copper, lead and zinc mining industry. Each mining area is treated separately, detailing the development of Colorado's mines on a county by county basis. **8.5" X 11", 284 ppgs, Retail Price: $19.99**

Gold Mining in Gilpin County Colorado - Unavailable since 1876, this publication was originally compiled by the Register Steam Printing House of Central City, Colorado. A rare glimpse at the gold mining history and early mines of Gilpin County, Colorado from their first discovery in the 1850's up to the "flush years" of the mid 1870's. Of particular interest is the history of the discovery of gold in Gilpin County and details about the men who made those first strikes. Special focus is given to the early gold mines and first mining districts of the area, many of which are not detailed in other books on Colorado's gold mining history. **8.5" X 11", 156 ppgs, Retail Price: $12.99**

Mining in the Gold Brick Mining District of Colorado - Important insights into the history of the Gold Brick Mining District, as well as its local geography and economic geology. Also included are the histories and locations of historic mines in this important Colorado Mining District, including the Cortland, Carter, Raymond, Gold Links, Sacramento, Bassick, Sandy Hook, Chronicle, Grand Prize, Chloride, Granite Mountain, Lucille, Gray Mountain, Hilltop, Maggie Mitchell, Silver Islet, Revenue, Roosevelt, Carbonate King and others. In addition to hardrock mining, are also included are details on gold placer mining in this portion of Colorado. **8.5" X 11", 140 ppgs, Retail Price: $12.99**

Ore Deposits of the London Fault of Colorado - First published in 1941, it has been unavailable since those days and sheds important light on the mines and mineral deposits of the London Fault in Central Colorado's Alma Mining District. This publication sheds important light on the gold veins and lead-silver deposits of the Alma Mining District. Included are geologic details on the London Mine, American Mine, Havigorst Tunnel, Ophir Mine, Mosher Tunnel, London-Butte Mine, Venture Shaft, Hard-To-Beat Mine, Oliver Twist Tunnel, Sacramento Mine, Mudsill Mine, Sherwood Mine, Wagner, Barcoe Tunnel and other mines in this important mining region. 110 ppgs., 10.99

The Mines of Colorado - First published in 1867, it has been unavailable since those days and sheds important light on Colorado's early mining history. Written shortly after the events took place, this publication sheds important light on the Pike's Peak Gold Rush, the discovery of gold on Ralston Creek and Dry Creek in the 1850's, as well as details on the first wave of miners into Colorado and their trials and tribulations as they crossed the Great Plains. Also included are details on early discoveries of lode gold in the mountainous regions of Colorado, details on the early mines hardrock and placer mines, and much more. It is a veritable treasure trove on Colorado's early mining history and will be of great importance to anyone who is interested in the mining of gold or other minerals in Colorado, as well as those interested in the history of the state. 478 ppgs., 29.99

The La Plata Mining District of Colorado - Originally titled "Geology and Ore Deposits in the Vicinity of the La Plata District of Colorado" and first published in 1949, it has been unavailable since those days and sheds important light on the mines and mineral deposits of the La Plata Mining District of Colorado. 214 ppgs., 19.99

Washington Mining Books

The Republic Mining District of Washington - Unavailable since 1910, this important publication was originally published by the Washington Geologic Survey and has been unavailable for a century. Topics include the geology, rock formations and the formation of ore deposits in this important mining area of Washington State. Also included are hard to find details on the geology, history and locations of dozens of mines in the area. Some of the mines featured include the New Republic Mine, Ben Hur, Morning Glory, the South Republic Mine, Quilp, Surprise, Black Tail, Lone Pine, San Poil, Mountain Lion, Tom Thumb, Elcaliph and many others. **8.5" X 11", 94 ppgs, Retail Price: $10.99**

The Myers Creek and Nighthawk Mining Districts of Washington - Unavailable since 1911, this important publication was originally published by the Washington Geologic Survey and has been unavailable for a century. Topics include the geology, rock formations and the formation of ore deposits in these important mining areas of Washington State. Also included are hard to find details on the geology, history and locations of dozens of mines in the area. Some of the mines featured include the Grant Mine, Monterey, Nip and Tuck, Myers Creek, Number Nine, Neutral, Rainbow, Aztec, Crystal Butte, Apex, Butcher Boy, Molson, Mad River, Olentangy, Delate, Kelsey, Golden Chariot, Okanogan, Ohio, Forty-Ninth Parallel, Nighthawk, Favorite, Little Chopaka, Summit, Number One, California, Peerless, Caaba, Prize Group, Ruby, Mountain Sheep, Golden Zone, Rich Bar, Similkameen, Kimberly, Triune, Hiawatha, Trinity, Hornsilver, Maquae, Bellevue, Bullfrog, Palmer Lake, Ivanhoe, Copper World and many others. **8.5" X 11", 136 ppgs, Retail Price: $12.99**

The Blewett Mining District of Washington - Unavailable since 1911, this important publication was originally published by the Washington Geologic Survey and has been unavailable for a century. Topics include the geology, rock formations and the formation of ore deposits in this important mining area of Washington State. Also included are hard to find details on the geology, history and locations of dozens of mines in the area. Some of the mines featured include the Washington Meteor, Alta Vista, Pole Pick, Blinn, North Star, Golden Eagle, Tip Top, Wilder, Golden Guinea, Lucky Queen, Blue Bell, Prospect, Homestake, Lone Rock, Johnson, and others. **8.5" X 11", 134 ppgs, Retail Price: $12.99**

Silver Mining In Washington - Unavailable since 1955, this important publication was originally published by the Washington Geologic Survey. Featured are the hard to find locations and details pertaining to Washington's silver mines. **8.5" X 11", 180 ppgs, Retail Price: $15.99**

The Mines of Snohomish County Washington - Unavailable since 1942, this important publication was originally published by the Washington Geologic Survey and has been unavailable for seventy years. Featured are details on a large number of gold, silver, copper, lead and other metallic mineral mines. Included are the locations of each historic mine, along with information on the commodity produced. **8.5" X 11", 98 ppgs, Retail Price: $10.99**

The Mines of Chelan County Washington - Unavailable since 1943, this important publication was originally published by the Washington Geologic Survey and has been unavailable for seventy years. Featured are details on a large number of gold, silver, copper, lead and other metallic mineral mines. Included are the locations of each historic mine, along with information on the commodity. **8.5" X 11", 88 ppgs, Retail Price: $9.99**

Metal Mines of Washington - Unavailable since 1921, this important publication was originally published by the Washington Geologic Survey and has been unavailable for nearly ninety years. Widely considered a masterpiece on the Washington Mining Industry, "Metal Mines of Washington" sheds light on the important details of Washington's early mining years. Featured are details on hundreds of gold, silver, copper, lead and other metallic mineral mines. Included are hard to find details on the mineral resources of this state, as well as the locations of historic mines. Lavishly illustrated with maps and historic photos and complete with a glossary to explain any technical terms found in the text, this is one of the most important works on mining in the State of Washington. No prospector or miner should be without it if they are interested in mining in Washington. **8.5" X 11", 396 ppgs, Retail Price: $24.99**

Gem Stones In Washington - Unavailable since 1949, this important publication was originally published by the Washington Geologic Survey and has been unavailable since first published. Included are details on where to find naturally occurring gem stones in the State of Washington, including quartz crystal, amethyst, smoky quartz, milky quartz, agates, bloodstone, carnelian, chert, flint, jasper, onyx, petrified wood, opal, fire opal, hyalite and others. **8.5" X 11", 54 ppgs, Retail Price: $8.99**

The Covada Mining District of Washington - Unavailable since 1913, this important publication was originally published by the Washington Geologic Survey and has been unavailable for a century. Topics include the geology, rock formations and the formation of ore deposits in this important mining area of Washington State. Also included are hard to find details on the geology, history and locations of dozens of mines in the area. Some of the mines featured include the Admiral, Advance, Algonkian, Big Bug, Big Chief, Big Joker, Black Hawk, Black Tail, Black Thorn, Captain, Cherokee Strip, Colorado, Dan Patch, Dead Shot, Etta, Good Ore, Greasy Run, Great Scott, Idora, IXL, Jay Bird, Kentucky Bell, King Solomon, Laurel, Laura S, Little Jay, Meteor, Neglected, Northern Light, Old Nell, Plymouth Rock, Polaris, Quandary, Reserve, Shoo Fly, Silver Plume, Three Pines, Vernie, White Rose and dozens of others. **8.5" X 11", 114 ppgs, Retail Price: $10.99**

The Index Mining District of Washington - Unavailable since 1912, this important publication was originally published by the Washington Geologic Survey and has been unavailable for a century. Topics include the geology, rock formations and the formation of ore deposits in this important mining area of Washington State. Also included are hard to find details on the geology, history and locations of dozens of mines in the area. Some of the mines featured include the Sunset, Non-Pareil, Ethel Consolidated, Kittaning, Merchant, Homestead, Co-operative, Lost Creek, Uncle Sam, Calumet, Florence-Rae, Bitter Creek, Index Peacock, Gunn Peak, Helena, North Star, Buckeye. Copper Bell, Red Cross and others. **8.5" X 11", 114 ppgs, Retail Price: $11.99**

Mining & Mineral Resources of Stevens County Washington - Unavailable since 1920, this important publication was originally published by the Washington Geologic Survey and has been unavailable for a century. Topics include the geology, rock formations and the formation of ore deposits in these important mining areas of Washington State. Also included are hard to find details on the geology, history and locations of hundreds of mines in the area. **8.5" X 11", 372 ppgs, Retail Price: $24.99**

The Mines and Geology of the Loomis Quadrangle Okanogan County, Washington - Unavailable since 1972, this important publication was originally published by the Washington Geologic Survey and has been unavailable for a century. Topics include the geology, rock formations and the formation of ore deposits in this important mining area of Washington State. Also included are hard to find details on the geology, history and locations of dozens of gold, copper, silver and other mines in the area. **8.5" X 11", 150 ppgs, Retail Price: $12.99**

The Conconully Mining District of Okanogan County Washington - Unavailable since 1973, this important publication was originally published by the Washington Geologic Survey and has been unavailable for a century. Topics include the geology, rock formations and the formation of ore deposits in this important mining area of Washington State, which also includes Salmon Creek, Blue Lake and Galena. Also included are hard to find details on the geology, mining history and locations of dozens of mines in the area. Some of the mines include Arlington, Fourth of July, Sonny Boy, First Thought, Last Chance, War Eagle-Peacock, Wheeler, Mohawk, Lone Star, Woo Loo Moo Loo, Keystone, Hughes, Plant-Callahan, Johnny Boy, Leuena, Gubser, John Arthur, Tough Nut, Homestake, Key and many others **8.5" X 11", 68 ppgs, Retail Price: $8.99**

Wyoming Mining Books

Mining in the Laramie Basin of Wyoming - Unavailable since 1909, this publication was originally compiled by the United States Department of Interior. Also included are insights into the mineralization and other characteristics of this important mining region, especially in regards to coal, limestone, gypsum, bentonite clay, cement, sand, clay and copper. **8.5" X 11", 104 ppgs, Retail Price: $11.99**

New Mexico Mining Books

The Mogollon Mining District of New Mexico - Unavailable since 1927, this important publication was originally published by the US Department of Interior and has been unavailable for 80 years. Topics include the geology, rock formations and the formation of ore deposits in this important mining area in New Mexico. Of particular focus is information on the history and production of the ore deposits in this area, their form and structure, vein filling, their paragenesis, origins and ore shoots, as well as oxidation and supergene enrichment. Also included are hard to find details, including the descriptions and locations of numerous gold, silver and other types of mines, including the Eureka, Pacific, South Alpine, Great Western, Enterprise, Buffalo, Mountain View, Floride, Gold Dust, Last Chance, Deadwood, Confidence, Maud S., Deep Down, Little Fanney, Trilby, Johnson, Alberta, Comet, Golden Eagle, Cooney, Queen, the Iron Crown, Eberle, Clifton, Andrew Jackson mine, Mascot and others. **8.5" X 11", 144 ppgs, Retail Price: $12.99**

The Percha Mining District of Kingston New Mexico - Unavailable since 1883, this important publication was originally published by the Kingston Tribune and has been unavailable for over one hundred and thirty five years. Having been written during the earliest years of gold and silver mining in the Percha Mining District, unlike other books on the subject, this work offers the unique perspective of having actually been written while the early mining history of this area was still being made. In fact, the work was written so early in the development of this area that many of the notable mines in the Percha District were less than a few years old and were still being operated by their original discoverers with the same enthusiasm as when they were first located. Included are hard to find details on the very earliest gold and silver mines of this important mining district near Kingston in Sierra County, New Mexico. **8.5" X 11", 68 ppgs, Retail Price: $9.99**

East Coast Mining Books

The Gold Fields of the Southern Appalachians - Unavailable since 1895, this important publication was originally published by the US Department of Interior and has been unavailable for nearly 120 years. Topics include the geology, rock formations and the formation of ore deposits in this important mining area of the American South. Of particular focus is information on the history and statistics of the ore deposits in this area, their form and structure and veins. Also included are details on the placer gold deposits of the region. The gold fields of the Georgian Belt, Carolinian Belt and the South Mountain Mining District of North Carolina are all treated in descriptive detail. Included are hard to find details, including the descriptions and locations of numerous gold mines in Georgia, North Carolina and elsewhere in the American South. Also included are details on the gold belts of the British Maritime Provinces and the Green Mountains. **8.5″ X 11″, 104 ppgs, Retail Price: $9.99**

Gold Rush Tales Series

Millions in Siskiyou County Gold - In this first volume of the "Gold Rush Tales" series, leading mining historian and editor Kerby Jackson, introduces us to the story of how millions of dollars worth of gold was discovered in Siskiyou County during the California Gold Rush. Lavishly illustrated with photos from the 19th Century, this hard to find information was first published in 1897 and sheds important light onto the gold rush era in Siskiyou County, California and the experiences of the men who dug for the gold and actually found it. **8.5″ X 11″, 82 ppgs, Retail Price: $9.99**

The California Rand in the Days of '49 - In this second volume of the "Gold Rush Tales" series, leading mining historian and editor Kerby Jackson, introduces us to four tales from the California Gold Rush. Lavishly illustrated with photos from the 19th Century, this hard to find information was first published in 1890's and includes the stories of "California's Rand", details about Chinese miners, how one early miner named Baker struck it rich and also the story of Alphonzo Bowers, who invented the first hydraulic gold dredge. **8.5″ X 11″, 54 ppgs, Retail Price: $9.99**

More Mining Books

Prospecting and Developing A Small Mine - Topics covered include the classification of varying ores, how to take a proper ore sample, the proper reduction of ore samples, alluvial sampling, how to understand geology as it is applied to prospecting and mining, prospecting procedures, methods of ore treatment, the application of drilling and blasting in a small mine and other topics that the small scale miner will find of benefit. **8.5″ X 11″, 112 ppgs, Retail Price: $11.99**

Timbering For Small Underground Mines - Topics covered include the selection of caps and posts, the treatment of mine timbers, how to install mine timbers, repairing damaged timbers, use of drift supports, headboards, squeeze sets, ore chute construction, mine cribbing, square set timbering methods, the use of steel and concrete sets and other topics that the small underground miner will find of benefit. This volume also includes twenty eight illustrations depicting the proper construction of mine timbering and support systems that greatly enhance the practical usability of the information contained in this small book. **8.5″ X 11″, 88 ppgs. Retail Price: $10.99**

Timbering and Mining - A classic mining publication on Hard Rock Mining by W.H. Storms. Unavailable since 1909, this rare publication provides an in depth look at American methods of underground mine timbering and mining methods. Topics include the selection and preservation of mine timbers, drifting and drift sets, driving in running ground, structural steel in mine workings, timbering drifts in gravel mines, timbering methods for driving shafts, positioning drill holes in shafts, timbering stations at shafts, drainage, mining large ore bodies by means of open cuts or by the "Glory Hole" system, stoping out ore in flat or low lying veins, use of the "Caving System", stoping in swelling ground, how to stope out large ore bodies, Square Set timbering on the Comstock and its modifications by California miners, the construction of ore chutes, stoping ore bodies by use of the "Block System", how to work dangerous ground, information on the "Delprat System" of stoping without mine timbers, construction and use of headframes and much more. This volume provides a reference into not only practical methods of mining and timbering that may be employed in narrow vein mining by small miners today, but also rare insights into how mines were being worked at the turn of the 19th Century. **8.5″ X 11″, 288 ppgs. Retail Price: $24.99**

A Study of Ore Deposits For The Practical Miner - Mining historian Kerby Jackson introduces us to a classic mining publication on ore deposits by J.P. Wallace. First published in 1908, it has been unavailable for over a century. Included are important insights into the properties of minerals and their identification, on the occurrence and origin of gold, on gold alloys, insights into gold bearing sulfides such as pyrites and arsenopyrites, on gold bearing vanadium, gold and silver tellurides, lead and mercury tellurides, on silver ores, platinum and iridium, mercury ores, copper ores, lead ores, zinc ores, iron ores, chromium ores, manganese ores, nickel ores, tin ores, tungsten ores and others. Also included are facts regarding rock forming minerals, their composition and occurrences, on igneous, sedimentary, metamorphic and intrusive rocks, as well as how they are geologically disturbed by dikes, flows and faults, as well as the effects of these geologic actions and why they are important to the miner. Written specifically with the common miner and prospector in mind, the book will help to unlock the earth's hidden wealth for you and is written in a simple and concise language that anyone can understand. **8.5″ X 11″, 366 ppgs. Retail Price: $24.99**

Mine Drainage - Unavailable since 1896, this rare publication provides an in depth look at American methods of underground mine drainage and mining pump systems. This volume provides a reference into not only practical methods of mining drainage that may be employed in narrow vein mining by small miners today, but also rare insights into how mines were being worked at the turn of the 19th Century. **8.5″ X 11″, 218 ppgs. Retail Price: $24.99**

Fire Assaying Gold, Silver and Lead Ores - Unavailable since 1907, this important publication was originally published by the Mining and Scientific Press and was designed to introduce miners and prospectors of gold, silver and lead to the art of fire assaying. Topics include the fire assaying of ores and products containing gold, silver and lead; the sampling and preparation of ore for an assay; care of the assay office, assay furnaces; crucibles and scorifiers; assay balances; metallic ores; scorification assays; cupelling; parting' crucible assays, the roasting of ores and more. This classic provides a time honored method of assaying put forward in a clear, concise and easy to understand language that will make it a benefit to even beginners. **8.5″ X 11″, 96 ppgs. Retail Price: $11.99**

Methods of Mine Timbering - Originally published in 1896, this important publication on mining engineering has not been available for nearly a century. Included are rare insights into historical methods of timbering structural support that were used in underground metal mines during the California that still have a practical application for the small scale hardrock miner of today. **8.5″ X 11″, 94 ppgs. Retail Price: $10.99**

The Enrichment of Copper Sulfide Ores - First published in 1913, it has been unavailable for over a century. Topics include the definition and types of ore enrichment, the oxidation of copper ores, the precipitation of metallic sulfides. Also included are the results of dozens of lab experiments pertaining to the enrichment of sulfide ores that will be of interest to the practical hard rock mine operator in his efforts to release the metallic bounty from his mine's ore. **8.5″ X 11″, 92 ppgs. Retail Price: $9.99**

A Study of Magmatic Sulfide Ores - Unavailable since 1914, this rare publication provides an in depth look at magmatic sulfide ores. Some of the topics included are the definition and classification of magmatic ores, descriptions of some magmatic sulfide ore deposits known at the time of publication including copper and nickel bearing pyrrohitic ore bodies, chalcopyrite-bornite deposits, pyritic deposits, magnetite-ileminite deposits, chromite deposits and magmatic iron ore deposits. Also included are details on how to recognize these types of ore deposits while prospecting for valuable hardrock minerals. **8.5″ X 11″, 138 ppgs. Retail Price: $11.99**

The Cyanide Process of Gold Recovery - Unavailable since 1894 and released under the name "The Cyanide Process: Its Practical Application and Economical Results", this rare publication provides an in depth look at the early use of cyanide leaching for gold recovery from hardrock mine ores. This volume provides a reference into the early development and use of cyanide leaching to recover gold. **8.5″ X 11″, 162 ppgs. Retail Price: $14.99**

California Gold Milling Practices - Unavailable since 1895 and released under the name "California Gold Practices", this rare publication provides an in depth look at early methods of milling used to reduce gold ores in California during the late 19th century. This volume provides a reference into the early development and use of milling equipment during the earliest years of the California Gold Rush up to the age of the Industrial Revolution. Much of the information still applies today and will be of use to small scale miners engaging in hardrock mining. **8.5″ X 11″, 104 ppgs. Retail Price: $10.99**

Leaching Gold and Silver Ores With The Plattner and Kiss Processes - Mining historian Kerby Jackson introduces us to a classic mining publication on the evaluation and examination of mines and prospects by C.H. Aaron. First published in 1881, it has been unavailable for over a century and sheds important light on the leaching of gold and silver ores with the Plattner and Kiss processes. **8.5″ X 11″, 204 ppgs. Retail Price: $15.99**

The Metallurgy of Lead and the Desilverization of Base Bullion - First published in 1896, it has been unavailable for over a century and sheds important light on the the recovery of silver from lead based ores. Some of the topics include the properties of lead and some of its compounds, lead ores such as galenite, anglesite, cerussite and others, the distribution of lead ores throughout the United States and the sampling and assaying of lead ores. Also covered is the metallurgical treatment of lead ores, as well as the desilverization of lead by the Pattinson Process and the Parkes Process. Hofman's text has long been considered one of the most important early works on the recovery of silver from lead based ores. **8.5" X 11", 452 ppgs. Retail Price: $29.99**

Ore Sampling For Small Scale Miners - First published in 1916, it has been unavailable for over a century and sheds important light on historic methods of ore sampling in hardrock mines. Topics include how to take correct ore samples and the conditions that affect sampling, such as their subdivision and uniformity. Particular detail is given to methods of hand sampling ore bodies by grab sample, pipe sample and coning, as well as sampling by mechanical methods. Also given are insights into the screening, drying and grinding processes to achieve the most consistent sample results and much more. **8.5" X 11", 124 ppgs. Retail Price: $12.99**

The Extraction of Silver, Copper and Tin from Ores - First published in 1896, it has been unavailable for over a century and sheds important light on how historic miners recovered silver, copper and tin from their mining operations. The book is split into three sections, including a discussion on the Lixiviation of Silver Ores, the mining and treatment of copper ores as practiced at Tharsis, Spain and the smelting of tin as it was practiced by metallurgists at Pulo Brani, Singapore. Also included is an overview and analysis of these historic metal recovery methods that will be of benefit to those interested in the extraction of silver, copper and tin from small mines. **8.5" X 11", 118 ppgs. Retail Price: $14.99**

The Roasting of Gold and Silver Ores - First published in 1880, it has been unavailable for over a century and sheds important light on how historic miners recovered gold and silver rom their mining operations. Topics include details on the most important silver and free milling gold ores, methods of desulphurization of ores, methods of deoxidation, the chlorination of ores, methods and details on roasting gold and silver ores, notes on furnaces and more. Also included are details on numerous methods of gold and silver recovery, including the Ottokar Hofman's Process, the Patera Process, Kiss Process, Augustin Process, Ziervogel Process and others. **8.5" X 11", 178 ppgs. Retail Price: $19.99**

The Examination of Mines and Prospects - First published in 1912, it has been unavailable for over a century and sheds important light on how to examine and evaluate hardrock mines, prospects and lode mining claims. Sections include Mining Examinations, Structural Geology, Structural Features of Ore Deposits, Primary Ores and their Distribution, Types of Primary Ore Deposits, Primary Ore Shoots, The Primary Alteration of Wall Rocks, Alterations by Surface Agencies, Residual Ores and their Distribution, Secondary Ores and Ore Shoots and Vein Outcrops. This hard to find information is a must for those who are interested in owning a mine or who already own a lode mining claim and wish to succeed at quartz mining. **8.5" X 11", 250 ppgs. Retail Price: $19.99**

Garnets: Their Mining, Milling and Utilization - First published in 1925, it has been unavailable since those days and sheds important light on the mining, milling and utilization of garnets. Included are details on the characteristics of garnets, where they are found and how they were mined. **78 ppgs, 10.99**

Gemstones and Precious Stones of North America - Leading mining historian Kerby Jackson introduces us to a classic mining publication on the gems and precious stones of the United States, Canada and mexico. First published in 1890, it has been unavailable since those days and sheds important light on the gems and precious stones that may be found in North America. Included are chapters on diamonds, corundum, sapphire, ruby, topaz, emerald, disapore, spinel, turquoise, tourmaline, garnets, beyrl, peridot, zircon, quartz crystals, feldspars, pearls and many others. Included are details on where these gems and precious stones may be found throughout North America, as well as their characteristics. **360 ppgs, 24.99**

Mining Camps and Mining Districts - First released in 1885 by Charles Howard Shinn under the title "Mining Camps: A Study in American Frontier Government", this publication offers a unique look at how early gold miners established their own forms of representative government during the California Gold Rush. Drawing on the the early mining codes of mideviel German miners in the Harz Mountains, on the mining customs of the Cornish tin miners and early Spanish mining laws introduced into California, the miners established the first governments in the American West. **340 ppgs, 24.99**

BLM Field Handbook for Mineral Examiners - Leading mining historian Kerby Jackson introduces us to a classic mining publication on mine evaluation. First published in 1962, this work sheds important light on the techniques of BLM Mineral Examiners to perform validity on mining claims. **132 ppgs, 10.99**

Six Months In The Gold Mines During The California Gold Rush - Unavailable since 1850, this important work is a first hand account of one "49'ers" personal experience during the great California Gold Rush, shedding important light on one of the most exciting periods in the history of not only California, but also the world. Compiled from journals written between 1847 and 1849 by E. Gould Buffum, a native of New York, "Six Months In The Gold Mines During The California Gold Rush" offers a rare look into the day to day lives of the people who came to California to work in her gold mines when the state was still a great frontier. **8.5" X 11", 290 ppgs. Retail Price: $19.99**

The Discovery of Gold in Australia - First published in 1852, it has been unavailable since those days and sheds important light on Australia's gold mining history. Included are rare communications between British agents and the British Crown when gold was first discovered in Australia in 1851. This rare text contains hard to find details on Australia's first mining camps and Britain's early attempts to provide for the orderly regulation of gold mines in that part of the world. Also of interest are hard to find extracts of articles that appeared in the early colonial newspapers that did their best to report on Australia's gold rush as it took place.
102 ppgs, 10.99

www.ingramcontent.com/pod-product-compliance
Lightning Source LLC
Chambersburg PA
CBHW080657190526
45169CB00006B/2157